ECOLOGICAL GOLF COURSE MANAGEMENT

By
Paul D. Sachs
and
Richard T. Luff

JOHN WILEY & SONS, INC.

This book is printed on acid-free paper. ♾

Published by John Wiley & Sons, Inc., Hoboken, New Jersey
Published simultaneously in Canada.

For general information on our other products and services or for technical support, please contact our Customer Care Department within the United States at (800)-762-2974, outside the United States at (317) 572-3993 or fax (317) 572-4002.

Wiley also publishes its books in a variety of electronic formats. Some content that appears in print may not be available in electronic books. For more information about Wiley products, visit our web site at www.wiley.cor

Library of Congress Cataloging-in-Publication Data:

Sachs, Paul D.
Ecological golf course management / Paul D. Sachs, Richard T. Luff.
p. cm.
Includes bibliographical references and index.
ISBN 1-57504-154-5
1. Golf courses—Environmental aspects. 2. Turf management—Environmental aspects. 3. Ecosystem management. I. Luff, Richard T. II. Title.
GV975.3.S33 2002
796.352'06'9—dc21 2001006101

Printed in the United States of America

10 9 8 7 6 5 4 3 2

To
Ruth Ann
and
Nancy-Jane

In memory of
Peter Luff

Acknowledgments

Special thanks to David Bergquist, Vicki Bess, Doug Brede, Will Brinton, Paul Chu, Charlie Clark, Peter Cookingham, Karl Danneberger, Dan Dinnelli, Jeff Frank, Jennifer Grant, Elizabeth Guertal, Wendy Sue Harper, Mary Howell-Martens, Norman Hummel, Elaine Ingham, Pat Kriksceonaitis, George Leidig, Barbara Luff, Eric Nelson, Martin Petrovic, Larry Phelon, Daniel Potter, Bill Quarles, James Ricci, Bruce Richards, Frank Rossi, Troy Russell, Christopher Schardl, Jim Snow, Ray Tennenbaum, Brian Vinchesi, Patricia Vittum, and Peter Wild for sharing their knowledge and experience with us.

About the Authors

Paul Sachs is the founder and owner of North Country Organics, a Bradford, Vermont based manufacturer and supplier of natural fertilizers, soil amendments, and environmentally compatible pest controls.

Paul has studied natural soil system dynamics for over 18 years and is considered one of the foremost authorities in the country on organic land care. He has written three books, hundreds of articles for trade journals, and speaks regularly at association conferences for professionals involved in both agriculture and horticulture. He has served as a member of the Technical Advisory Panel for the National Organics Standards Board of the United States Department of Agriculture. He also works as a consultant for a highly diversified group of clients. In 1993 he completed work on his first book entitled *EDAPHOS: Dynamics of a Natural Soil System*, in 1996 finished his second book, entitled *Handbook of $uccessful Ecological Lawn Care*, and in 1999 he completed a second edition of *EDAPHOS: Dynamics of a Natural Soil System.*

Richard Luff is former superintendent and presently general manager of the Sagamore Hampton Golf Club, a course that has been maintained ecologically since 1962. He is intimately familiar with ecological maintenance procedures on the golf course and has successfully controlled most turfgrass pests biologically, botanically, or with cultural practices. His familiarity with golf course construction and maintenance dates back three generations. His father, R. DeWitt Luff, introduced him to ecological maintenance procedures as an adolescent. Richard also is a consultant for other courses that are trying to reduce or eliminate pesticides.

Contents

Preface

The keynote speaker at the 2000 Long Island Organic Horticulture Association conference was evangelical as he spoke about the relationship between quantum physics and organic horticulture. His enthusiasm radiated out over the audience like a broadband signal from a powerful transmitter but, for the most part, the attendees' antennae were down. With youthful excitement and the passion of a gospel preacher, the orator spoke of ancient thoughts that are as true today as they were over 2000 years ago, but many in this audience were not receiving the message. They didn't grasp the holistic significance of all things, big and small, animate and inanimate, performing in an orchestrated system of life. They did not accept the transcendental truths handed down through generations of agrarians who survived through participation as opposed to arrogation. They could not believe that even the most brilliant scientists on earth are completely inept and unqualified to redesign nature and that we, the magnificent human race, are humbled by the genius and complexity of our ecosystem. Many of the listeners just sat and sighed with a glazed look in their eyes, waiting patiently for the speaker to tell them about the four-step program that would magically transform the relatively uncomplicated task of applying chemicals to an equally unchallenging job of putting down 'earth friendly' materials.

"Rocks think!" the speaker cried, his outstretched hands reverberating with his words. The sighs increased from attendees nearby; the glazed eyes were harder to see, as many were now closed. The largely blue-collar audience came to learn how they could conform to the eco-mores being forced upon them by people terrified of an unseen but empirically evident toxic monster. They wanted to know what to do, what to use, how to use it, and when to use it. They didn't come to understand the philosophies of men who died before Christ was born. They didn't want to know the relationships between matter and energy, life and death, animate and inanimate, or quantum physics and organic horticulture. They wanted directions.

Their attitude was without apathy, arrogance, or ignorance and it was not condescending or intolerant; it was just what one should expect. For the past 50 years these people had been taught that nothing is true unless it is proven by science—that the wisdom of a hundred generations was nothing more than folklore and, consequently, lost and forgotten arts. At the dawn of the scientific era, science became the new religion and scientists the priests and preachers of this gospel. Indeed, to some of the more arrogant scientists who proselytized *better living through chemistry*, nature was nothing more than a big pain in the ass that needed to be tamed, not understood. This audience wanted and needed to hear that same confident and somewhat arrogant science that they had been hearing for the past five decades. They came to learn of some new, natural strategy that would work everywhere, all the time. This is not what they heard.

Between the thinking rocks, the cosmic particles passing through lead without collisions, and the guardian spirits protecting every living thing on earth, the speaker had some very pertinent and important messages. The panacea of all horticultural ills does not exist. There is no single strategy, either chemical or natural, that will work everywhere, all the time. Every site is as unique as a fingerprint. How can one strategy work everywhere? His was a message about respect—for ageless wisdom and for life, the most valuable component of our ecosystem. It was about faith, that nature can provide a system capable of sustaining life—our lives as well as all others. He told his audience we need to provide for nature, more accurately for nature's life forms, and they will provide for us. This cannot be done with biocides that are designed to end life. It was also about faith in ourselves—that we are capable of cultivating a productive relationship with the other organisms in our ecosystem without resorting to chemical warfare.

Faith was the biggest hurdle for many in this audience. They had been taught, from the beginning, that the scientific method is the only acceptable way wisdom can become knowledge. If it can't be proved or explained, it has little validity. Asking this audience to have faith in a universal truth was as difficult as convincing an atheist of salvation. Many of the attendees did not seem ready to make this quantum leap of faith.

The information in this book is not as cerebral as what the speaker was trying to convey. It doesn't compare quantum physics with ecological golf course management. However, it doesn't present a four-step program that works as well in New York as it does in Arizona, either. The notion of such a program is preposterous. Most of the information presented in the following pages is knowledge substantiated by scientific research, but anecdotal and empirical evidence is also discussed. The experiences of superintendents and other turf managers is often as valid—if not more valid than research, and certainly worthy of mention.

Having respect for the ecosystem that sustains our turf is a message that reverberates throughout this book. We wholeheartedly believe that the system nature provides cannot be improved upon but it can, to a certain degree, be persuaded to help us produce a better landscape. We are not suggesting that prayer is the solution—although there is no research that suggests it will cause harm. The ecosystem in which our turf is grown depends on life, death, organic matter, inorganic minerals, atmosphere, solar radiation, and probably quantum physics to function. The life in a healthy soil is a powerful ally that needs to be nourished. When resources dwindle, so do the benefits these organisms provide. Without their help, the need for chemical support begins to increase along with the resultant danger of dependency.

In the early days of golf course maintenance, *good lie* was a major concern. Before tools were available to mow and apply fertilizer, animals were called upon to complete both tasks, often simultaneously. This natural system had its benefits, but soil-dwelling organisms and plants, not golfers, enjoyed most of them. Many of the problems faced by modern superintendents were not a concern back then. Weeds were plants that the animals wouldn't eat; turf disease was hardly ever mentioned. The playing surface left by hoofed mowers that chewed rather than cut was by today's standards pretty rough. Even if a golfer made an excellent drive straight down the fairway, it was anybody's guess as to which way it would bounce or roll. And even if the bounce and roll were somewhat true, it was an anxious moment for the golfer approaching the ball, wondering if it had *good lie*.

In the early days when golfers played with a feathery (a leather ball stuffed with feathers), the ball was probably light enough and perhaps big enough to guarantee better lie, but the feathery didn't carry very far. As the density of golf balls increased and the size decreased, they could be hit much farther, but it took a very skilled golfer to overcome the natural obstacles common to the golf courses of yesteryear. Often it was impossible to make a shot if the ball had very bad lie, and the rules of St. Andrews did not make allowances to save the situation. The differences between *good lie* and *bad lie* are much more subtle today—at least on the fairways of a modern American golf course.

In the early days of mechanized maintenance, most of the equipment used was agricultural. In fact, during the nineteenth century, many courses in the British Isles were on farmland that was harvested or grazed during the growing season and borrowed for golf in the fall and winter. Fairways and greens were cut with horse-drawn sickle-bar mowers and the greens were smoothed with large iron or wooden rollers, also drawn by horses. The roughs were cut once or twice a year if at all. In the mid 1800s, gang-type reel mowers were introduced but were still towed by horses. The horses' hooves were clad with wide-soled leather boots to mitigate the indentation they made in the soil. Horses were ubiquitous then and their manure was the logical choice for fertilizer. From a perspective of sustainability, this was an excellent system, but the golfer had his qualms.

For as many years as there have been golfers there have been complaints about the condition of the course. But it's reasonable to assume that as course conditions improved, the frequency of complaints increased. It's unlikely that all courses or even parts of courses improved simultaneously. So if a player tasted the forbidden fruit of a smooth, low-cut fairway or green on the first hole but not on the second, third, and fifth, or enjoyed these amenities at another course, the home course superintendent was sure to hear about it. When all holes on all courses were consistently poor, expectations were probably low along with the number of complaints.

These complaints, of course, were the driving force in the evolution of maintenance procedures. Not only did golf courses want to avoid losing members to rival clubs but they also wanted to attract new members. To do that they needed to produce courses that favored less-skilled golfers. Greens and fairways had to offer flawless lie, true bounce, and extensive rollout and, at the same time, be aesthetically pleasing. Weeds, bare spots, or the discoloration from dehydration were unacceptable. No longer was play the only consideration. When a member chances upon a course groomed as eye-candy, his home course super will surely hear about the bright yellow dandelion in the middle of the fifth fairway.

The greater the number of players who are attracted to the game, the greater the number of complaints, usually blaming the course's conditions for the golfer's apparent lack of skill. There is, of course, a relationship between the number of rounds played and the number of wear-related problems, but many experienced, dedicated superintendents have felt the golfer's wrath because of aesthetic issues that do not affect the game.

Of all the changes that have occurred in golf courses over the centuries, construction techniques and the evolution of equipment have probably been the most dramatic, and the vast improvements to playing surfaces are mostly attributable to these technological advances. Greens, fairways, and even roughs are so different from those existing a century ago that, aside from the rules of St. Andrews, the game is entirely different. Greens are probably four to five times faster than they used to be and *bad lie* on the fairway, or even the rough, is relatively uncommon. Unfortunately, these wonderful conditions are not so terrific from the perspective of the turf plant. A healthy grass plant mowed at an eighth of an inch is like a healthy maple tree with 95% of its limbs removed. And the maple tree is not usually growing in almost pure sand. The sand-based green may be an improvement for the golfer, rendering a hard, smooth, well-drained surface but, again, it is not advantageous from the perspective of the plant.

Given the stressful and adverse conditions under which the plant is asked to grow on a golf course green, it's easy to understand how health problems can develop. These problems can be exacerbated by the deficient biological activity that is common in a sandy environment where there is usually an inadequate supply of resources for soil organisms. Health problems weaken turf and present opportunities for pathogens, insects, and weeds to succeed. Horticultural scientists recommend symptom relievers (pesticides) to deal with these problems, but as more and more of these products are applied, a chemical dependence is often established. What many of us involved in turf maintenance fail to embrace is that disease, weeds, and insects are not the cause of unhealthy turf—they result from it.

In the old world of turfgrass maintenance, before such an extensive chemical arsenal was available, keeping the grass as healthy as possible was the first line of defense against problems. Much attention was paid to soil fertility and structure, and organic matter was considered essential. Most of what is published about golf course maintenance, no matter how long ago it was written, proclaims plant health as an important step toward preventing pest problems. Many books and articles, however, especially those written in the 1950s and 1960s, recommended the newest and most powerful pesticide at the slightest hint of a problem. Indeed, many recommended the precautionary use of pesticides in anticipation of a problem that did not yet exist.

Now the green industry has come full circle. The huge chemical arsenal we once had at our disposal is slowly but surely being restricted from use. Either by local, regional, state, or even federal regulation or outright banning, fewer of these tools are available to us. The green industry is fighting hard to retain the right to use these chemicals but the trend is against them. The chemical industry is responding with stronger active ingredients to reduce application rates and with shorter environmental duration to mitigate long-term ecological impact and shorten reentry periods. Unfortunately, if the effect of these bioicides doesn't last as long, it's inevitable that they will have to be used more often. Realistically speaking, most pest controls don't solve problems, they just mitigate or eradicate the symptoms. Revelations that date back to Rachel Carson indict many pesticides as the cause of other problems that include stress-related incidences of plant disease, insect infestations, and weed invasion, not to mention pollution and human health problems.

What we propose in this book is to partially or perhaps completely liberate the modern superintendent from this chemical dependence by suggesting methods that don't adversely affect the game or the appearance of the course but dramatically reduce the need for pesticides and other chemicals. The political battle against the systematic reduction of chemical choices may only postpone the inevitable. On the other hand, learning alternative and ecological methods of golf course management can be an intriguing pursuit with tremendous rewards. But it is as much an art as it is a science. An art is a skill gained from experience, observation, or both. It cannot always be replicated or proven by the unskilled, inexperienced, or unobservant. Techniques that work for one superintendent but not for another are not necessarily invalid. They are either specific to the site where they are successful or they require a degree of skill and finesse that isn't being applied elsewhere.

The information in this book is not a formula for every golf course on earth. It is a presentation of alternatives and information that will enable the superintendent to consider or even invent different ways of solving problems that may include using some botanical or biological biocides, changing cultural practices, or just cultivating a healthier and more biologically active ecosystem. Maintaining a golf course ecologically means the superintendent should consider all of the living organisms that exist in the golf course environment because most of them are allies. It also means that expectations may have to be adjusted to a more realistic and practical threshold where a natural equilibrium can be maintained. It doesn't mean that beauty or playability have to be sacrificed.

AUTHORS' NOTE: Jeff Frank (the keynote speaker mentioned above) runs a school of environmental horticulture called the Lyceum and although we may have made Jeff sound a bit eccentric, we applaud his efforts and success in educating hundreds of professionals on how to use more ecological methods of maintaining beautiful landscapes. (See Sources and Resources at the back of this book for contact information.)

Chapter 1:

The Soil Ecosystem

If one were to contemplate the soil ecosystem as a machine, it would have billions of functioning parts. It would be such a complex mechanism that, were it to break, there would be no repairman on earth capable of fixing it. Fortunately, in its exquisite design, the soil also contains its own repair mechanism and even in cases of extreme damage, it can almost always repair itself. The time that it takes to implement recovery depends on the extent of the damage and it depends on conditions that regulate the renewal of life, because it is the soil's biological component that facilitates its ability to heal.

In 1980, Mount Saint Helens exploded and destroyed a 234 square mile area surrounding the volcano. The soil disappeared under a thick blanket of volcanic ash and rock. Soil life—all life—was annihilated. The devastation from extremely hot ash, pyroclastic flows, blowdowns, and mudflows was severe and seemingly irreparable but slowly, over the past two decades, life has begun to reestablish itself. The repair is an ongoing, joint effort between plants, soil organisms, insects and other arthropods, amphibians, birds, and mammals, some of whom work 24/7/365. Restoration probably began by photosynthesizing (autotrophic) bacteria that could extract energy from the sun, combine it with carbon dioxide from the atmosphere, and produce organic materials necessary for proliferation. As these pioneers of life grew in numbers, some predator organisms were able to subsist and they in turn fed others on the next link of the food chain. As more and more groups of different organisms began to colonize the devastated area, organic residues from their activities and expired bodies began to accumulate. Before long, plants were able to acquire the necessities for life. Wildflowers, especially prairie lupines, were probably the first to reappear. The prairie lupine survives and thrives in nutrient poor soils and, because it's a legume, the lupine fixes nitrogen (with the help of rhizobacteria) from the atmosphere for itself and shares it with other organisms. As more and more different plants appeared, wildlife could begin to graze again. Gophers were some of the first mammals to migrate into the devastation and their burrowing activities, through the ash to the soil below, began mixing the different horizons together. The spread and proliferation of life begun by autotrophic bacteria have, in the past 20 years, brought about significant change to an area that, a relatively short time ago, resembled the surface of the moon more than anything on this planet. As more and more life inhabits this devastated soil, it becomes more inhabitable for other organisms including forest plants. The essential interdependence of life in a functioning ecosystem is an important lesson to learn from Mount Saint Helens and one that applies everywhere.

The chemical, biological, and physical reactions that make soil functional need air,

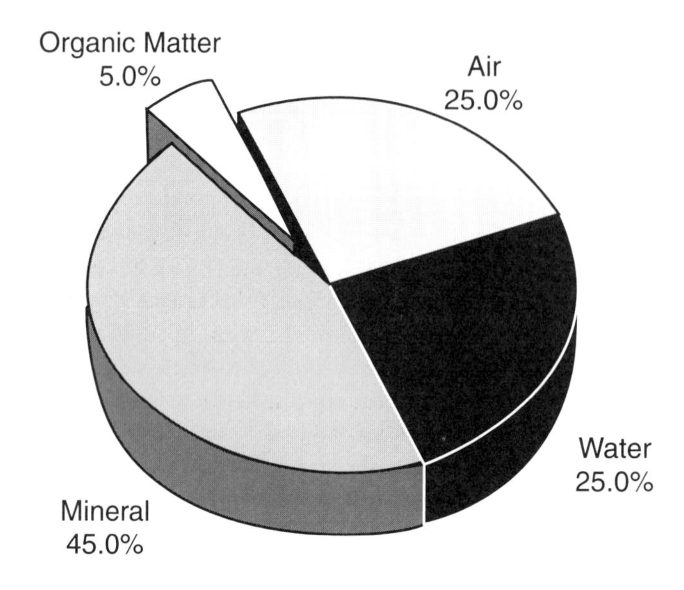

water, minerals, energy from the sun, and time. The scenic beauty of natural landscapes was not created from a grand plan engineered by a landscape architect and constructed by men in a single season. It evolved over thousands of millennia from organisms in stiff competition but, at the same time, creating symbiotic alliances along the way. It is that competition and those alliances that need to be better understood to engage the soil ecosystem in the creation of a sustainable recreational landscape. The soil machine can be nudged perhaps, but it cannot be controlled, which is fortunate because no one really understands the complexities of the soil system well enough to govern it. When control is attempted, a part of the machine may be altered or damaged and cease to function properly. In a simple machine, like an automobile engine, malfunctioning parts are noticed almost immediately. But in a very complex system, like the soil, dysfunction can go unnoticed for years, decades, or perhaps even centuries. And, because the machine is so complex, the cause of a symptom can easily be misunderstood.

To get a sense of how the soil functions—and it will be a superficial sense at best—let's first take a look at how the soil, as we know it, was formed.

Before there was soil, there was rock—perhaps one extremely large chunk called earth. Eventually, tectonic movement, volcanoes, and other natural forces created pieces of rock, some of which were very large and some were particles of dust. This process of rock size reduction is called weathering and there are many forces that contribute to it. The typical analysis of a well-developed loam (see Figure 1.1) shows that half of the soil's volume is pore space that is (ideally) filled with equal parts of air and water. Most of the soil solids are minerals derived from rock.

Water, especially when it freezes, is a bull in a china shop when it comes to weathering. In a river, stream, or brook, water constantly washes away surface particles from rocks in its path as it cascades over, under, around, and sometimes through the parent material. What may often seem gentle to the observer is an unrelenting torrent of force to the particles clinging to the rock's surface. Those particles snatched by the water's flow become unwitting accomplices at the tasks downstream, abrasively betraying brethren pieces. When gravity's assistance wanes, the water slows and many of the suspended particles settle to the bottom. Eventually the bottom rises from a buildup of these deposits and the water finds a new path, leaving behind beds of sediment—the foundation of a riparian soil.

When the temperature drops below freezing, water changes from liquid to solid and expands with a force that few natural materials can contain. One hundred fifty tons of expansive force per ft^2 can compromise the structure of nearly any rock that allows moisture to enter through cracks, fissures, or pores. If the force of frost separates a boulder from its mother mountain, gravity can assist further in the process of weathering as the accelerating rock smashes itself and the surface against which it impacts into smaller and smaller pieces.

Wind is another persuasive natural force that not only can coax small particles of rock away from its parent's surface but also spread it to areas far and wide. As these pieces ride the wind, they too become unwitting accomplices, blasting free other particles with which they collide.

Glaciers, earthquakes, and volcanic eruptions are other forces that produce weathered rock particles. As powerful and persuasive as all these physical forces can be, there is another type of weathering that also makes significant contributions to the formation of soil.

Surprisingly, the largest facility on earth where chemical reactions occur is the natural environment. Naturally occurring elements react with each other on a regular basis to form compounds, many of which then react with other compounds or elements. This constant manufacture and disintegration of chemicals in nature is an integral part of terrestrial functions. One of the effects of this natural chemical activity is an advanced stage of weathering rock into soil.

Water is chemically expressed as H_2O. Those two elements (hydrogen and oxygen) can chemically react with many other elements in nature. The entire water molecule can react in a process called *hydration*. Rock minerals that bond with one or more water molecules become hydrated and are more easily dissolved into a soil system. *Hydrolysis* occurs when a hydrogen atom in water bonds with natural elements, often forming acids that can further weather rock surfaces. Chemical chain reactions, initiated by either the hydrogen or the oxygen in water, can change the original composition of rock; the resulting mineral compounds can have completely different structures and reactive characteristics within the soil.

The formation of inorganic acids, such as sulfuric or hydrochloric, occurs naturally through reactions between different soil chemicals. These acids are extremely effective at separating and dissolving rock components. Carbonic acid, another effective weathering

FIGURE 1.2. CLAY PARTICLES

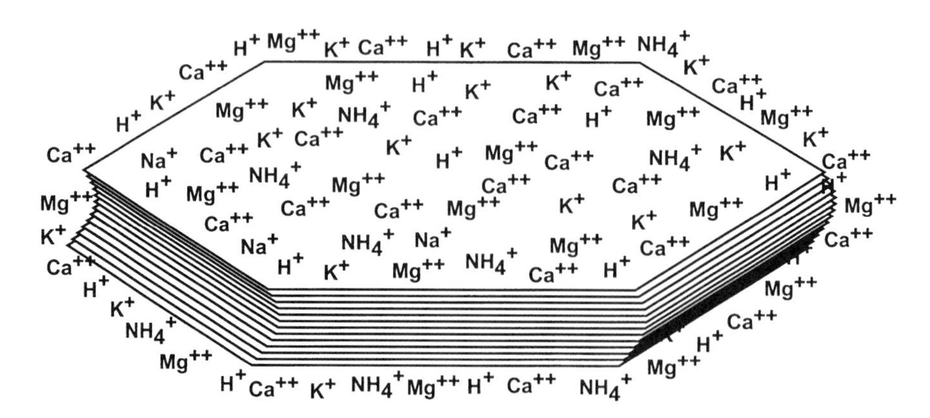

Negatively charged clay particles shown with typical plate-like appearance and swarm of adsorbed cations

agent, is formed from the combination of carbon dioxide and water, two relatively abundant substances in the soil.

The different types of parent material determine the rate at which rock is weathered and, to a large extent, the different sized particles found in soil. Limestone, for example, is a rock that is easily weathered and can eventually dissolve so completely that very few particles can be found. Quartz, on the other hand, weathers very slowly and because of its structure, it is difficult for nature to completely weather it into its molecular components.

Surface area is another factor that influences weathering. The greater the surface area exposed to soil acids, the faster rock can be dissolved. A fist-sized stone may have several square inches of surface area when intact, but when ground into a fine powder the overall surface area increases to several acres. The amount of time it takes for nature to weather the material would be measured in years for the powder, as opposed to centuries for the intact stone.

Rocks comprised predominately of aluminum, potassium, or magnesium silicates (generally insoluble compounds) are commonly weathered into very small (<0.002 inch diameter) plate-like particles classified as clay. Clay particles, because of complex substitutions of elements within their molecular structure, inherently have a negative electromagnetic (anionic) charge which enables them to adsorb positively charged ions (cations) such as potassium (K), calcium (Ca), or magnesium (Mg) (see Figure 1.2). This magnetic ability is described as colloidal (from the Greek *koll* meaning glue and *oid* meaning like) and is inherent in humus particles as well. Colloidal particles play a crucial role in the soil system. Soil environments devoid of either clay or humus—like pure sand greens—have a greatly diminished capacity to hold many plant nutrients and, consequently, do not naturally support an abundance of plants or other biological life.

Soils formed from rock are called mineral soils; they are the most common types of soil on earth. An analysis of a well-developed mineral soil might contain around 90%

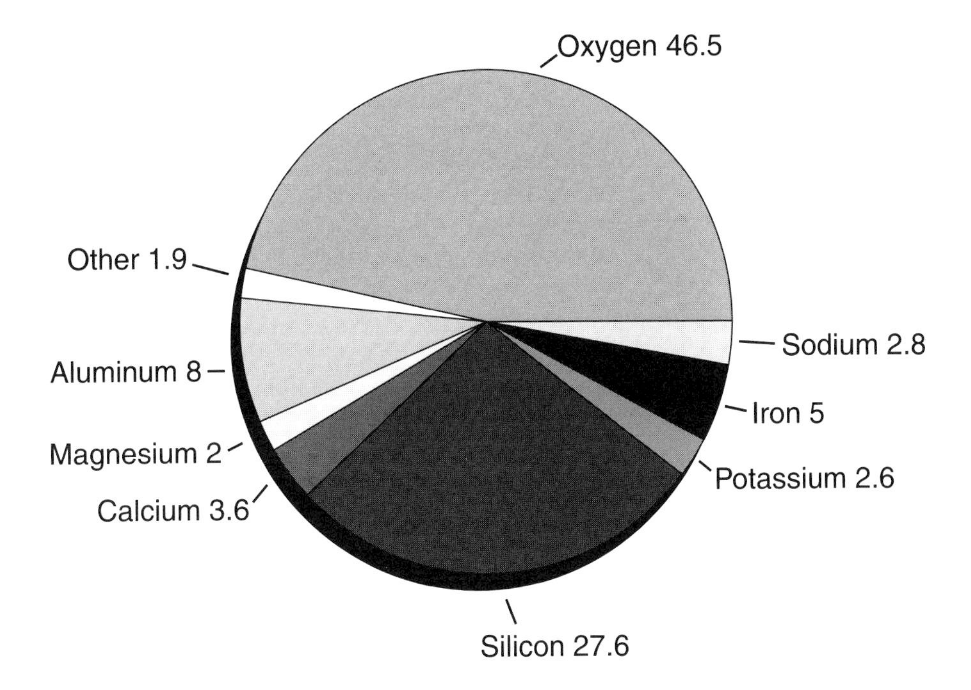

rock particles on a dry basis (see Figure 1.1). Volumetrically, 50%of this type soil will be made up of air and water. If it is a rich, healthy soil, an average of only 5% will be organic matter, although natural levels of organic matter vary considerably.

There are about 90 naturally occurring elements on earth; most are found, at least in trace amounts, nearly everywhere. The most common elements found in the mineral component of soil, however, are oxygen, silicon, aluminum, calcium, sodium, iron, potassium, and magnesium (see Figure 1.3). The oxygen that exists in soil minerals is part of the chemical structure and is not in a gaseous state. Many elements exist in a naturally formed molecule with oxygen.

All of these weathering forces, both physical and chemical, combine to form soil particles from rock that are better known as sand, silt, and clay. These particles are defined by size as shown in Table 1.1. Most soils have a combination of all three sizes of particles and are classified differently, depending on the percentage of each (see Chapter 4 for information on soil texture analysis).

The concept of weathering isn't difficult to grasp and seems to give a clear picture of how soil is formed, but it's not soil yet—just dirt. Weathering just provides a picture of how soil texture is formed.

Dirt needs something else before it can become soil, and that is life—and also death. Life and death in the soil are vital phenomena of an ecosystem that perpetually generate energy for the biosphere. Here is where the soil and its system of cycles begin to get more complex. Magdoff and van Es (2000) classified the organic fraction of the soil into three

Table 1.1. USDA Soil Particle Classification

Clay	≤ 0.002 mm
Silt	> 0.002 mm ≤ 0.05 mm
Sand	< 0.05 mm ≤ 2.0 mm
Gravel	> 2.0 mm

different categories, the living, the dead, and the very dead. The soil's living component includes the biomass—that group of organisms from the single-celled bacillus to macroorganisms, such as earthworms, arthropods, and mammals that live in the soil. But all organisms, including plants and humans, are connected to the soil in many ways. They are affected by and have effects on the soil. So all terrestrial life on earth should be included in this category.

Leaves that fall from trees, grass clippings, animals that burrow in or just trespass on the soil leave residues that contribute to the system of cycles. Most of these residues fall into the dead category. These residues not only provide energy and sustenance for numerous organisms but also contribute to the development of humus—the very dead.

Humus is a by-product and end product manufactured by organisms during decay processes. Humus is a dark, inconsistently shaped substance that is biologically resistant to decay and makes up the major portion of organic matter in most soils. It is a vital component of soil and provides a cornucopia of benefits.

Humus may be crucial to the existence of every living thing on earth but it is not a sexy topic of discussion. It constantly contributes to the mechanisms of life but can't carry on an engaging conversation or take out the garbage (but it has some friends that can break down food waste into a soil-like substance). It can't keep you warm at night unless you belong to a family of thermophilic bacteria and, although all material wealth in the world is linked either directly or indirectly with humus, it can't buy you a new car or anything else. It is important, however, to understand the role humus plays in the soil ecosystem. So put the kids to bed, grab a cup of strong coffee, and turn off the TV. This section may be as boring as the department of motor vehicle's driver education booklet but you needed to get your driver's license and you need to know this too. We guarantee this information is more interesting than the stuff you have to read to get (and keep) your pesticide applicator's license.

The discussion of humus that follows includes two distinctly different forms of humus. The first is young or friable humus and the second is stable humus. Friable humus is like compost. It has a wealth of resources available for soil organisms but is fragile and relatively ephemeral. Friable humus is still undergoing humification and very little of it may ever become stable humus, depending on the chemical structures of its organic contents and environmental conditions. Stable humus has much fewer available resources for soil organisms but makes greater contributions to soil structure and cation exchange capacity (CEC) than friable humus. Both friable and stable humus contribute to the soils' capacity to hold water and air—two essential constituents for most soil organisms (including turf plants). Both can increase a soil's resistance to compaction. In fact, humus is an amazing soil conditioner. Only 3 to 5% humus can transform almost lifeless sand into a relatively rich loam. It has abilities to both bind sand and flocculate clay. Plants tend to produce more microscopic root hairs when growing in a soil environment that contains adequate amounts of humus. These tiny root hairs have significantly more surface area than the coarse roots that typically grow in sandy soils and can absorb more water and nutrients than an equal mass of larger roots.

In sandy soils—those often prescribed for golf course construction—plant and microbial mucilage from humus can reduce the size of the pore space between sand particles, increasing the moisture holding capacity of the soil and reducing the leaching of soil solution with all the dissolved nutrients it carries. As the moisture holding capacity increases, more plants and soil organisms can inhabit the environment. This, in turn, promotes the creation of more humus. Under ideal conditions, the advancement of humus in sand eventually will develop the most preferred type of loam for plant production. Almost everyone associated with the design and maintenance of golf courses will tell you that the accumulation of organic matter is one of the last things a superintendent wants. The common belief is that left unchecked, organic matter will accumulate to monstrous levels that arrest drainage and weaken turf. In a near-sterile environment where only undecomposed organic residues can accumulate, this may very well happen. But in a biologically active environment where residues are constantly being consumed by soil organisms, this is rarely the case. Conditions for the development of humus in sand are seldom ideal. In tropical and subtropical environments, for example, where moisture, temperature, and the duration of the warm season are optimum for populations of bacteria, fungi, and other saprophytic (decay) organisms, organic matter is quickly assimilated back into the biomass. This rapid assimilation, coupled with the abundance of oxygen in porous sand, makes it extremely difficult, if not impossible for humus to accumulate to an adequate, let alone an extreme level. Unlike a rain forest, a golf course situated in a tropical or subtropical environment has little chance of maintaining adequate levels of humus mainly because not much organic debris is contributed to the soil. The types of debris that contribute to thatch are not what's being discussed here. These tough, fibrous residues may eventually contribute to the formation of humus but must be balanced with residues containing sugars, starches, and proteins such as clippings, organic fertilizers, or mature compost. Without this balance, not only is there little chance of soil improvement from humus accumulation, but also less carbon dioxide can be recycled back to the grass plants for the production of photosynthesized materials. Most golf course architects insist on sand-based greens, tees, and often fairways too for the water-draining characteristics; however, this near-obsession with drainage is obscuring our view of other necessary components of the plant-growing system. It's a little like washing and waxing a car twice a week but never changing the oil.

In clay soil, humus forms an alliance with clay particles. Complexes are formed between the two particles because both particles are colloidal, possessing an electronegative charge capable of attracting and holding cation nutrients. These complexes not only increase the soil's overall cation exchange capacity (CEC), but also mitigate the cohesive nature of clay by causing granulation and increase the decay resistance of humus giving it a longer life span. Clay on the course is considered an even more taboo soil material than organic matter. An abundance of clay is, without argument, a less than ideal medium on the course but does that mean there should be no clay in the soil? We know arsenic can be lethal but we also know that a little is essential to human health.

Colloidal refers to the attraction certain soil particles have for ions of mineral nutrients. Whether organic (humus) or mineral (clay), the colloid is a very small soil particle, often referred to as a *micelle* (meaning micro-cell), and carries a negative electromagnetic charge that can hold positively charged ion nutrients (cations), such as calcium, magnesium, ammonium, and potassium, in a manner that allows plant roots access to them. This phenomenon is called *cation exchange*. The quantity of clay and humus in a given soil is usually relative to the amount of cations the soil can retain but the type of clay and age of humus are also factors. Different types of clay have varying abilities to hold cation nutrients, as does humus at different stages of its development. The combined competence of clay and humus quantitatively determine the soil's capacity to hold cations, which is measured as CEC. Clay is not a material that is often used in golf course construction; however, many experiments have shown that small additions of materials like bentonite montmorillonite or zeolite significantly benefit the dynamics of a soil ecosystem. Not only do they increase CEC but they also improve plant performance.

The accumulation of humus is naturally easier in a soil containing clay than in sand because the environmental conditions for decay bacteria in the former tend to be less ideal. Moisture levels in clay soils will often reach the saturation point, leaving little room for the oxygen needed by aerobic life. Soil water also acts as a buffer for temperature changes, which often suppresses the warmth needed for biological activity. In addition, the evaporation of moisture from the surface has a cooling effect on the soil (just as evaporation of perspiration from the skin cools the body). Clay can also assist in the stabilization of humus. The clay-humus complexes formed in the soil can further inhibit bacterial decomposition and can increase the life span of humus to over a thousand years. Soil scientists calculate that in Allophanic soils (a volcanic clay soil) the residence time of humus ranges from 2,000 to 5,000 years. This clay-humus complex may form in some old fairways built on native soils but would be extremely rare in greens.

The formation of humus begins when residues from plants and animals come in contact with decay organisms (saprophytes) in an average soil. Much of the carbon compounds contained in those residues are proteins, carbohydrates, and other organic materials that supply energy to soil bacteria, fungi, actinomycetes, nematodes, earthworms, protozoa, and other organisms involved directly or indirectly in the decay process. Aerobic saprophytes are the most adept at decomposing organic matter. To live and be active, they need an environment where there is an adequate amount of free oxygen. The degree to which free oxygen exists in the soil plays a major role in regulating the favorable or unfavorable conditions under which humus is formed. The same is true for the amount of moisture, the soil temperature, and the carbon to nitrogen ratio of the residues being decomposed (see Chapter 3: Compost).

Where little or no free oxygen exists (for example, in stagnant water), decomposition of organic matter occurs mostly by anaerobic organisms. The process is slower than that conducted by aerobic organisms but can, in the long run, produce a greater amount of humus (as muck or organic soil). Okay! We understand that turf is not an aquatic plant but it's important that we present the whole picture, so bear with us. Humus formed under water is slightly different than its aerobic counterpart, due as much to the nature of the residues from the two different environments as to the process of aerobic vs. anaerobic humification. Most of the contributions of organic matter to organic soils are from water-dwelling insects and other organisms, which have a higher percentage of protein than do the plant residues found on a golf course. Other components come from organic residues transported by wind and water currents to a location where they can accumulate and settle. In fact, some of this translocated material may already be humus. Higher percentages of humus are found in soils formed anaerobically, because conditions are more favorable for humus accumulation and less favorable for its destruction.

At the other extreme is an environment where there is too much oxygen. If moisture and soil temperature are also at optimum levels, organic residues can be decomposed so quickly that little or no accumulation of humus occurs at all. We see this in tropical and subtropical environments, where high temperatures and moisture levels occur in predominantly sandy soils that naturally contain an abundance of air. Conditions such as these occur on many golf course greens, even in the more temperate regions of the

FIGURE 1.4. INFLUENCE OF TEMPERATURE ON SOIL ORGANIC MATTER CONTENT

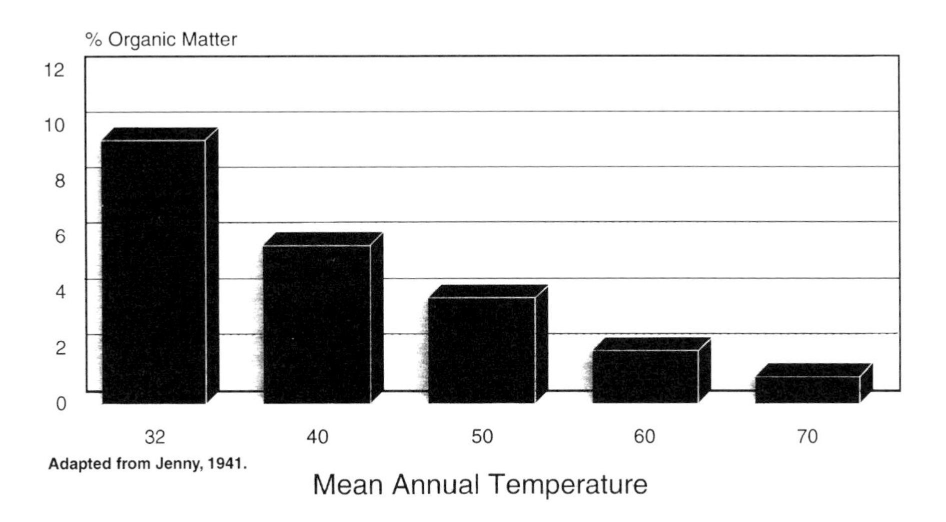

world. The soil is comprised mainly of sand, the turf is mowed too low to protect the soil from the sun's heat, and irrigation provides tropical rain forest-like precipitation.

Soil temperature is an important controlling factor involved in the formation of humus. As the temperature of a soil increases, there is a corresponding increase in biological activity. Soils that exist in warmer regions of the earth tend to have lower average levels of humus than soils in colder areas (see Figure 1.4). Figure 1.5 shows that at a soil temperature of 88° F (30° C), with adequate aeration, humus can no longer accumulate. Not only are the activities of decay organisms stimulated and prolonged but plants often

FIGURE 1.5. INFLUENCE OF SOIL TEMPERATURE ON HUMUS ACCUMULATION

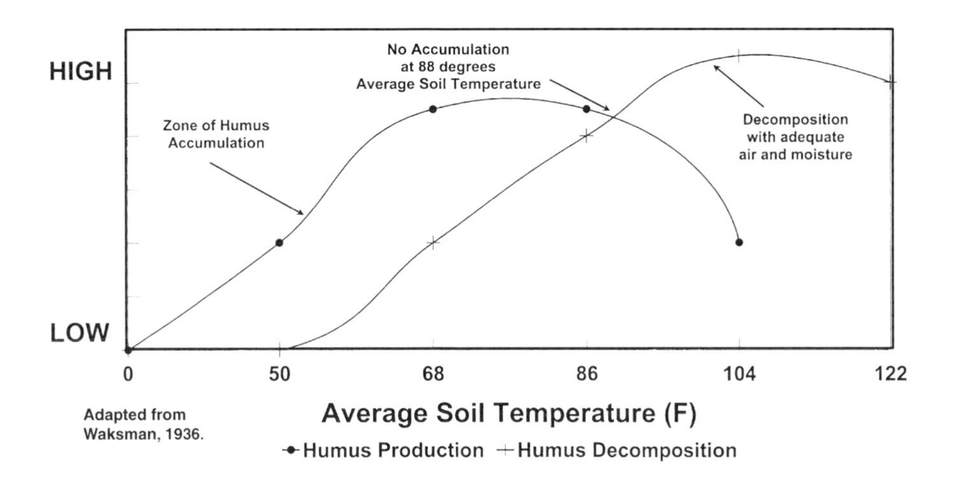

use more energy than they create when soil temperatures rise beyond a certain point. This phenomenon significantly reduces the amount of plant residues contributed to the soil and, because of the height of cut (HOC) on greens, tees, and even fairways, it occurs commonly on golf courses.

During humification of organic matter, saprophytes use most of the sugars, starches, proteins, cellulose, and other carbon compounds for their own metabolisms. The assimilation of nutrients and energy from organic residues is the first stage in the process of creating humus. It is not precisely known to what degree the different residues contribute to the creation of humus. Some of the more easily digested components of the residues end up being used by many different organisms, and may never actually become humus. However, these components contribute to the overall process by providing energy and protein for the life cycles of the organisms involved in the synthesis of humus. The components of the residues, which are more decay resistant, are not so much assimilated as they are altered by biological processing into humic substances. Materials produced by soil organisms involved in the humification process also contribute to the components of humus.

The nutrients and energy assimilated into the bodies of soil organisms is normally reused by other organisms that are either predacious or saprophytic. Some is assimilated by the consumer, some is mineralized into plant nutrients, and some is changed into biologically resistant compounds that accumulate as components of humus. The cycles of soil life are implemented as more and more members of the biomass club participate in the festivities of eating, dying, and being eaten—dead or alive. Plants create organic residues, which feed soil organisms that then transform the resources from the residues back into plant nutrients, into food for other organisms, and into humus.

FIGURE 1.6. HUMUS
Typical Consistency in Mineral Soils

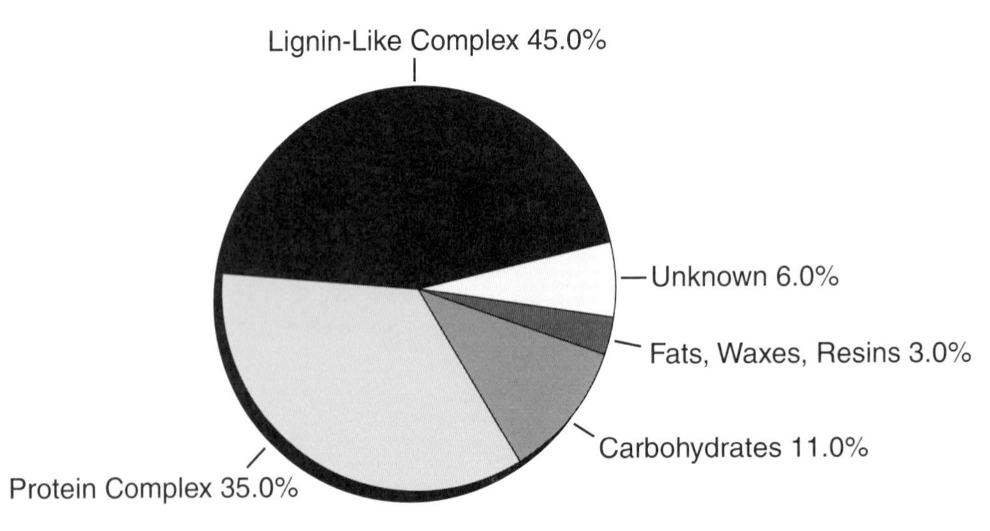

Adapted from Waksman, 1936.

Some of the components in organic residues are much more resistant to decay than others. Most carbohydrates, such as sugars and starches, decompose faster than materials such as cellulose. Of all organic components, fats, waxes, and lignin are the most resistant to decay. Proteins vary in decay resistance: they are generally more resistant than sugars and starches, but more easily decomposed than many other components.

Although many of these components exist in humus in a biologically altered form (see Figure 1.6), the degree to which they exist in the organic residues plays a role in the accumulation of humus. Materials that contain high percentages of easily decomposed components such as sugars, starches, and some proteins are, for the most part, assimilated back into the living biomass. Although the energy and protein provided by these residues help in the creation of humus, the ratio of their mass to the measure of humus produced is relatively wide (i.e., a very small amount of humus is created).

Materials that contain a large percentage of lignin, cellulose, or other biologically resistant components have less recyclable nutrients to offer the soil food web but contribute significantly more to the formation of humus. Different plant residues have inherently different ratios of these organic components, but variance also appears in the same plants at different stages of their lives. Green leaves from deciduous trees, for example, have a very different analysis of proteins vs. other components than do their dry, fallen counterparts. These same differences would appear in grass plants if they were allowed to mature and produce seed. The constant cutting of grass clippings, however, produces a fairly consistent residue throughout the growing season. At this young, succulent stage, grass clippings contribute far more recyclable nutrients and less of the substances needed for the accumulation of humus. However, much of the turf root system, which dies off at the end of every season, contributes many components needed for the production of humus. Thatch is also comprised of components that can contribute to the accumulation of humus but, for the most part, it doesn't. Unlike roots, thatch has limited contact with the soil and the decay organisms therein. It also cannot sustain moist enough conditions to support abundant soil life. Additionally, thatch residues lack sufficient sugars, starches, and proteins needed for rapid decay.

Biological processing structurally and chemically changes decay-resistant components such as lignin, fats, and waxes into other biologically resistant by-products and these decay-resistant compounds are what make up humus. This is not to say that humus is immune from further decay but its resistance to decomposition enables it to exist for decades, if not centuries, as a soil conditioner, a habitat for soil life, and a vast reservoir of nutrients and moisture for plants and many different groups of soil organisms.

The digestion of organic matter in the soil is analogous to the digestive system in mammals. Mammals ingest food and derive nutrients from it. The nutrients are diffused into the body, and used for energy to function and produce new cells. By-products, such as urea, water, carbon dioxide and other simple compounds, are given off. The indigestible portion of the food is excreted as feces.

In the soil, organisms assimilate organic residues, using the nutrients and energy for their own metabolisms. Their activities convert some of the organically bound nutrients back into a mineral form, which is usable by plants and other soil organisms. The indigestible portions of the residues accumulate as humus. However, humus is not completely immune to decomposition. Soil organisms will eventually recycle all the elements in humus, even if it takes a millennium (or longer).

Humus, plant residues, and other components of soil organic matter are like storage batteries containing energy originally derived from the sun. Researchers in England cal-

FIGURE 1.7. ENERGY FLOW

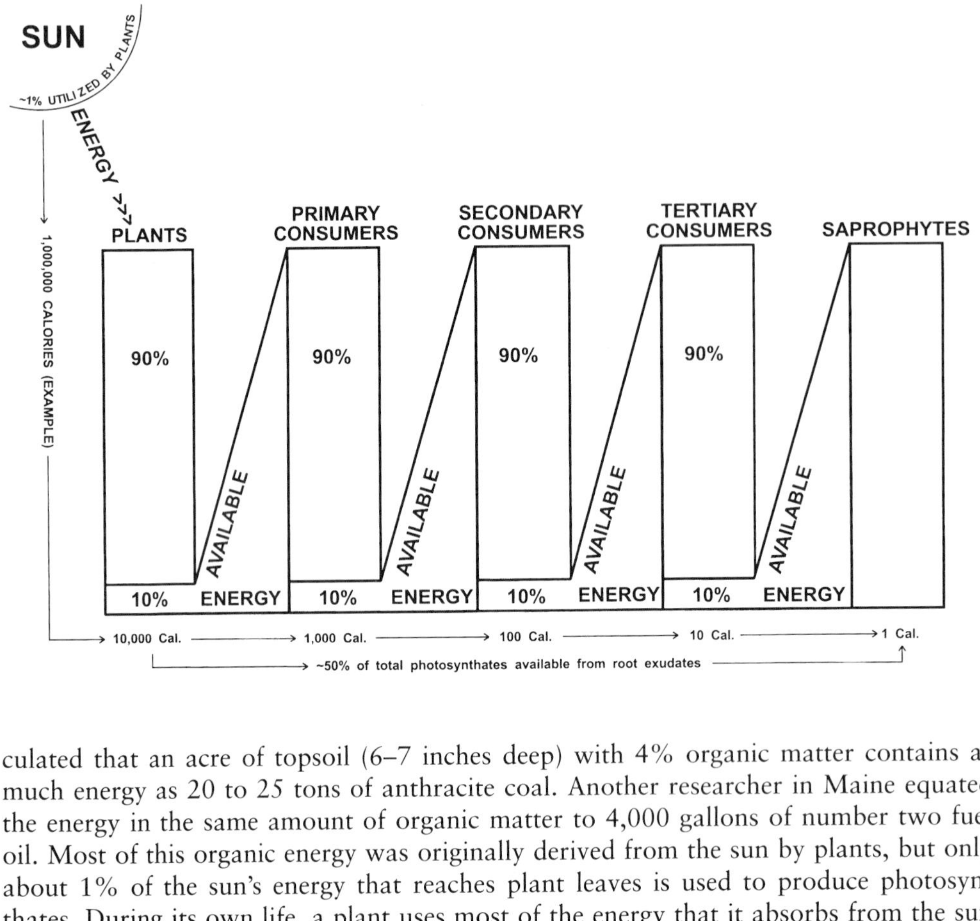

culated that an acre of topsoil (6–7 inches deep) with 4% organic matter contains as much energy as 20 to 25 tons of anthracite coal. Another researcher in Maine equated the energy in the same amount of organic matter to 4,000 gallons of number two fuel oil. Most of this organic energy was originally derived from the sun by plants, but only about 1% of the sun's energy that reaches plant leaves is used to produce photosynthates. During its own life, a plant uses most of the energy that it absorbs from the sun for growth, foliage production, flowering, seed production, and other functions. About 10% of the absorbed energy is left in the plant tissue for a consumer (e.g., an organism that eats and digests the plant). This leftover energy is called *net primary production.* Like the plant, the herbivore uses most of the energy it consumes for functions such as growth and sustenance, and offers about 10% of the energy it derived from plants to the next consumer in the food chain.

Subsequent digestions through the food chain continue the rapid depletion of available energy from one link to the next. Often the final consumers of this energy reside in the soil. In Figure 1.7, an arbitrary quantity of energy has been used as an example to show its flow and use. In this case, a figure of one million calories of energy offered by the sun is reduced to one calorie of available energy by the time it flows through the food chain to soil saprophytes. During the season when plants are active, however, up to half of their photosynthesized nutrients are released through the roots into the soil. Organisms living near, on, or within the root surface then use these compounds for sustenance. One might consider plants to be rather inefficient organisms for allowing half of their photosynthetic production to leak from their root systems, but nature has a good reason for this design. This phenomenon provides a direct and constant flow of plant-

FIGURE 1.8. CARBON CYCLE

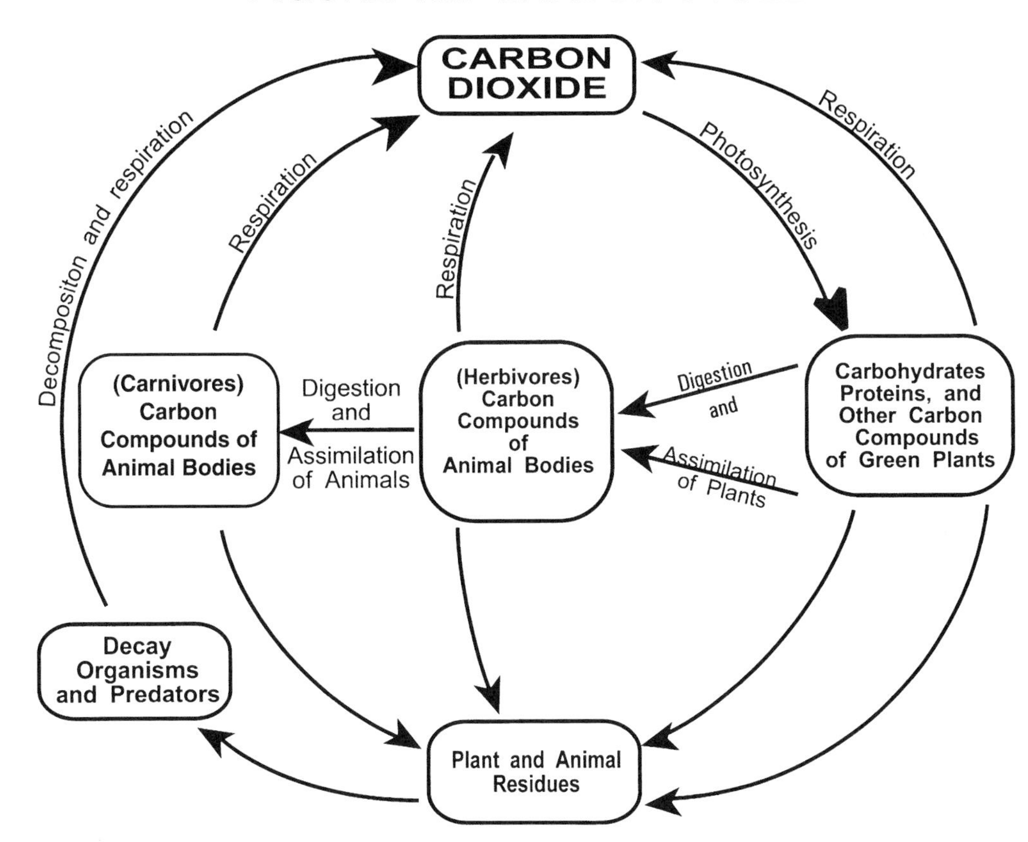

synthesized energy for many groups of beneficial soil organisms. The substrate exuded from the plant roots provides sustenance for huge populations of organisms that, among other things, make mineral nutrients available to the plants, protect the plants from pathogens and other pest organisms, and produce carbon dioxide that plants need to conduct more photosynthesis.

The amount of available energy in plant residues, in the remains of herbivores, and what is left over from carnivores is significantly different. The various energy levels of these different residues can stimulate different populations of soil organisms that can perform different functions in the soil. Their populations are often controlled by the amount and type of residues introduced into the soil which, in turn, control the quantity and characteristics of humus.

Throughout this digestion and assimilation process, from the first consumption of the sun's energy to the decomposition of all residues in the soil, carbon is released back into the atmosphere as carbon dioxide through a process called *respiration*. The evolution of carbon dioxide from the decay of organic residues is an integral part of the life cycle. Figure 1.8 shows how carbon from the atmosphere cycles through the food chain and back into the atmosphere. Plants need atmospheric carbon dioxide to live. If carbon dioxide were not evolved, it would not be available to plants and the accumulation of humus and other organic materials would bury the planet. Life, as we know it, could not exist.

Over a one-year period and under average conditions, about 60 to 70 % of the carbon in fresh organic residues is recycled back to the atmosphere as carbon dioxide. Five to ten percent is assimilated into the biomass and the rest resides in friable humus. Friable humus is not stable humus, however. It can take decades for humus to develop the biological resistances necessary to be considered stable.

We can see from the carbon cycle that it is necessary for humus to not only accumulate, but to be destroyed as well, so that carbon dioxide can eventually be returned to the atmosphere where plants can access it. In an uncultivated, natural environment, humus accumulates in accordance with the favorable and unfavorable conditions of the region. Unless global or regional conditions change, the level of humus accumulation reaches equilibrium with the factors that destroy it. The level of humus then becomes a relatively fixed component of that environment so long as the quantity and source of the organic residues being contributed to the soil remains relatively constant.

In cultivated environments, both friable and stable humus are important assets which, like most other assets, are easier to maintain than replace. Unfortunately, the value of humus is too often overlooked until it is severely depleted and its benefits are no longer available. Even then, deficient levels of organic matter are rarely recognized as a possible cause of problems that inevitably arise. The importance of soil organic matter and the populations of organisms that it supports should rank very high on the priority list at an ecological golf course. The first step toward preserving and maintaining this resource is to understand its value before it's gone.

Old, stable humus is biologically resistant. Depending upon the environmental conditions under which it exists, some humus can sit in the soil for centuries, even millennia, with only a minimal amount of decomposition. However slight, decay still occurs, and even old humus will eventually cycle back from where it came. The formation of new humus, which requires a constant source of organic residues, is critical to maintaining a stable presence of this asset.

Figure 1.9 shows a typical fate of organic matter introduced into the soil. It is important to note that even under the best conditions, a relatively small amount of humus is created in comparison to the level of organic residues initially introduced. If conditions exist that further accelerate the decomposition of organic matter, even less humus may eventually be created. In extremes, such as tropical or subtropical environments where moisture, heat, and soil oxygen are abundant, a great amount of carbon dioxide may be evolved, but there is little production of humus.

Organisms that create humus and those that prosper in soil enriched with organic matter also contribute to the phenomenon of soil formation and the availability of plant nutrients. Populations of soil organisms facilitate a degree of chemical weathering by creating many different organic acids—corrosive materials formed in nature by plants, animals, and soil organisms. They include citric acid, acetic acid, amino acid, lactic acid, salicylic acid, tannic acid, nucleic acid, and humic acid, to name only a few. All of these acids have varying abilities to react with rock surfaces and liberate minerals from parent material.

Soil organisms dissolve minerals from rock through three basic functions: *assimilation,* the creation of *organic acids,* and the creation of *inorganic acids.*

Assimilation is the direct absorption of nutrients by organisms. Most soil organisms work to free nutrients from organic and inorganic matter for their own metabolism. Saprophytes primarily feed on dead residues of plants and animals but also have the ability to absorb essential minerals from various inorganic sources in the soil. Predators feed on saprophytes and other organisms (including other predators) and much of what becomes available to plants is released during predation. Some organisms work together

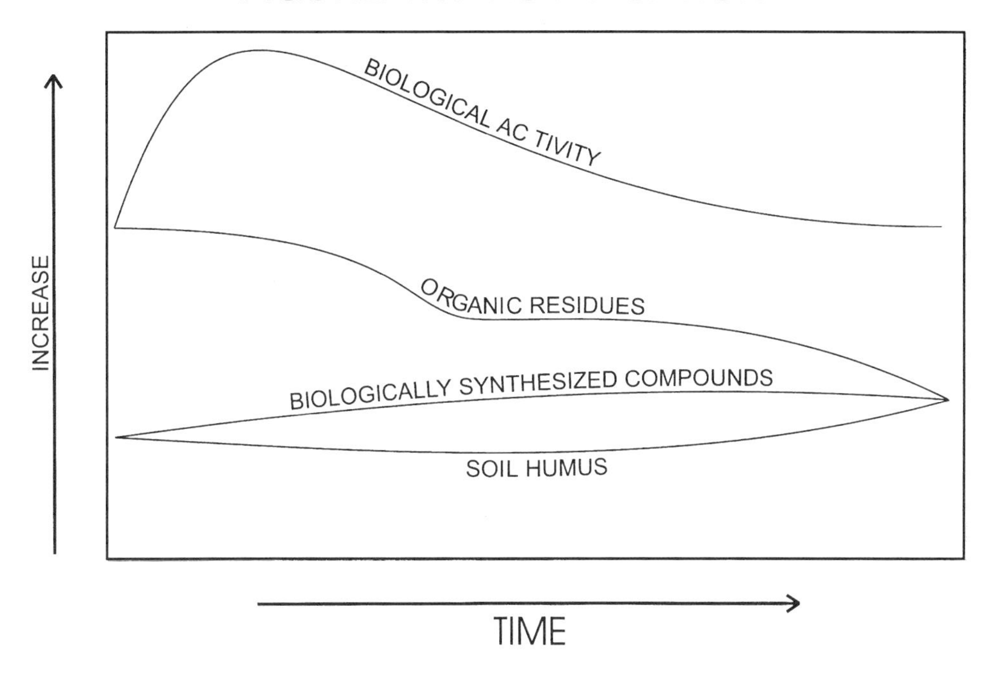

symbiotically to secure their nutrition. Lichen, for example, are symbiotic combinations of fungi and algae which attach themselves to various surfaces (often rocks) and dislocate minerals for their own nutrition. The microscopic hyphae of the fungal component can penetrate the finest hairline cracks in certain rock, releasing enzymes and acids powerful enough to liberate mineral ions from surrounding surfaces. Eventually, the minerals assimilated by lichen are recycled through the agency of other organisms that either feed on lichen or decompose their remains after death. After mineral nutrients are incorporated into the body of an organism, however, the nutrients are bound in organic compounds, which are easier for nature to process. Unless these elements are removed from a given area (by cropping, erosion, or other means) they tend to accumulate over time and increase fertility for both plants and other indigenous organisms.

Most organisms create enzymes and organic acids during assimilation or the acids are by-products of their metabolism. Regardless of their origin, these acids and enzymes contribute to the formation of soil and to the availability of many essential plant nutrients. Figure 1.10 shows the influence of Azotobacter bacteria on the mineral content in soil solutions from their organic acids. These beneficial organisms also fix nitrogen from the atmosphere.

Inorganic acids, such as sulfuric and nitric are formed in the soil as an indirect result of biological activity. Specific organisms such as sulfur bacteria or nitrifying bacteria oxidize (combine with oxygen) sulfur and nitrogen, respectively, which can then combine with soil hydrogen to form strong inorganic acids. The chemical reaction between rock surfaces and strong acids changes both substances. Part of the rock dissolves and the acid can then become a mineral salt, which is more soluble and, once dissolved, is more easily assimilated by plants and other soil organisms.

Mycorrhiza is a family of fungi that can dissolve mineral nutrient from rock surfaces. These organisms work under the soil's surface in a symbiotic relationship with perennial

FIGURE 1.10. INFLUENCE OF ACIDITY CREATED BY AZOTOBACTER ON SOLUTIONS OF CA, MG, AND K IN THE SOIL

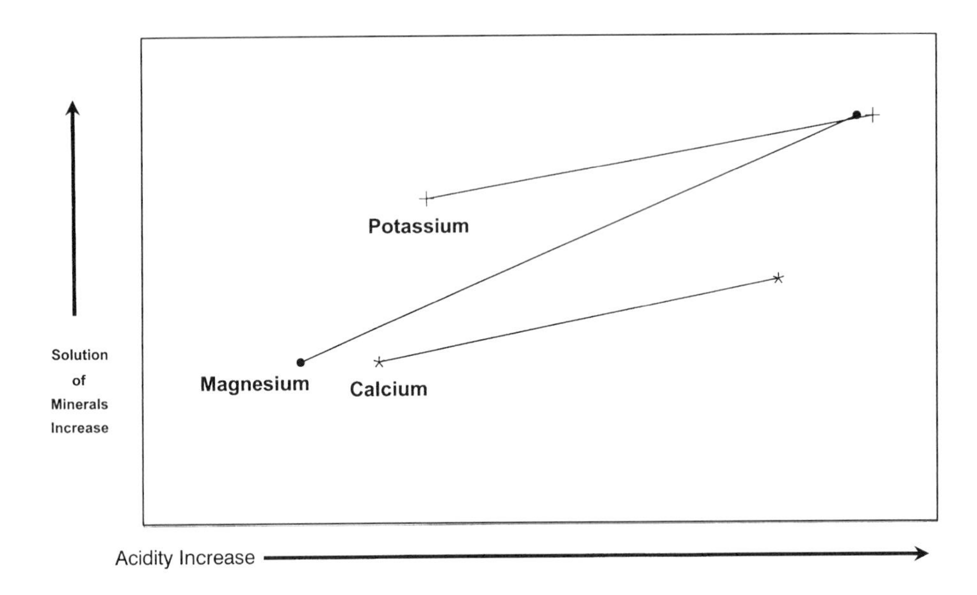

plant roots. The fungi attach themselves to roots and, in exchange for a small amount of carbohydrate supplied by the plant, these organisms provide water and mineral nutrients that they retrieve and transport via their hyphae (long, microscopic tubing) to the plant from regions of the soil beyond the reach of plant roots.

The chemical weathering effect that roots have is from several sources. Carbon dioxide given off by root hairs can combine with water to form carbonic acid. Research has shown significant mineral release from carbon dioxide when dissolved in water. This process is known as carbonation and involves the attachment of mineral ions, such as potassium, magnesium, or calcium, to a carbonate ion, thus forming other mineral salts. Plant roots indirectly participate in the formation of organic acids, because of their relationship with soil organisms. The population of organisms within the rhizosphere (the soil region immediately surrounding the root hair) is always significantly higher than in other parts of the soil. The production of organic acids and enzymes that are destructive to rock is directly proportional to the level of biological activity. Up to 50% of the carbon that plants fix from atmospheric carbon dioxide is released through their roots as carbohydrates, most of which are consumed by soil organisms that are constantly producing organic acids.

Plants are the link between the atmosphere and the pedosphere (heaven and earth). Their remains make rich soil from weathered rock particles, and they provide nutrients and energy for most other living things on the planet. They complete the cycles of water, nutrients, and energy. Plants are the producers. They combine energy from the sun, carbon dioxide in the atmosphere, and mineral nutrients in the soil to synthesize sustenance either directly or indirectly for almost every other living thing on earth. Almost all of the protein, energy, and carbohydrates that humans need come from plants or from animals raised on plants.

Plant life slowly began to appear as prehistoric soils evolved. The meager amount of available nutrients, coupled with what was probably a harsh environment permitted only small, perhaps microscopic, plants to grow at first. As time went on, the accumulation of organic residues from the remains of countless generations of microplant life slowly enriched the soil, creating an environment that could support a larger and more diverse population of soil organisms. The cycle snowballed for a period of time, creating a richer soil with each successive increase in vegetative establishment, until an ecological equilibrium was reached. At this point, the production of organic matter by plants is roughly equal to the environmental factors that destroy it.

Plants stabilized much of the volatility in the ecosystem. In a sense, plants bottled the energy of the sun and made it available to other living things. The evolution of plants and other photosynthetic organisms changed the ecosystem of the planet dramatically. Prior to autotrophs (photosynthesizing organisms), life existed on available energy. When the energy was depleted, the organisms often became extinct and new organisms evolved to recycle the energy left behind by their predecessors. The renewability of energy on earth enabled many species of consumers (heterotrophs, such as mammals) to exist for many generations without depleting their food supply. This phenomenon allowed for the systematic evolution of different species into present-day life forms.

Most plants are composed of two fundamentally different parts: the root system, which anchors the plant and absorbs water and mineral nutrients; and the shoot system which consists of the plant's trunk, branches, shoots, stems, and leaves. Those parts of the plant that connect the roots to the leaves provide structural support, as well as transport passageways for water and nutrients. Leaf surfaces contain pores, called *stomata* or *stomates*, that function as gas exchangers and act as part of the plant's ventilation system.

The plant's link to the atmosphere is through its use of carbon dioxide gas, with which it creates almost all the organic carbon compounds that exist on earth. Without carbon dioxide, plants could not live, and neither could any of the organisms that depend on the nutrients that plants provide. The balance of carbon dioxide with other gases in the atmosphere depends largely on plants or, more accurately, on the use of energy stored in plants. Whenever energy is extracted from plant tissue (or animal tissue), whether by animals, insects, or soil organisms, carbon dioxide is given off. Plants capture some carbon dioxide as it wafts from the soil's surface into the atmosphere and it is transformed into plant tissue. The more carbon dioxide that is generated by organisms under the plant, the more likely it is that adequate amounts will be captured by plant stomates located on the underside of the leaves.

Plant roots make the organic–inorganic connection in the soil. The fine root hairs combine so well with the earth that it is difficult to tell exactly where the soil ends and the root begins. The roots of plants depend on the shoot system for nourishment. Throughout the plant's life, carbohydrates and other organic nutrients flow down to the root system from the leaves, where energy from the sun, carbon dioxide from the atmosphere, and minerals drawn from the soil combine in a complex process called *photosynthesis*.

The root system and plant tops act like different organisms living together symbiotically. The leaves are autotrophic, using carbon dioxide from the atmosphere to manufacture organic carbon compounds, and giving off oxygen as a waste product. The root system acts like a heterotrophic organism, dependent upon the sugars and other compounds produced in the leaves, and using oxygen from the soil, while giving off carbon dioxide.

The roots of plants are more than just anchors and siphon tubes extracting water and nutrients from the soil. The rhizosphere (the zone of proximity around each root hair) is teeming with life in the form of billions of microorganisms. Up to 50% of the nutrients photosynthesized in plant leaves are released from the roots into the rhizosphere, supplying nourishment for most of these organisms. The activities of these organisms help the plant in numerous ways. They can mineralize organic nutrients, compete antagonistically with pathogens, weather parent material, regulate nutrient availability, decay plant residues, return carbon dioxide to the atmosphere, and mobilize water and nutrients (e.g., via mycorrhizae). Some organisms in the rhizosphere can also fix nitrogen from the atmosphere, solubilize phosphorus, and produce growth hormones that increase the plants' resistance to environmental and other types of stress.

Plants also affect soil factors that influence plant growth. Environments like prairies or forests are initially formed by other factors, such as climate, parent material, and topography, but the establishment of certain types of plants soon becomes another factor acting on the environment. The plant factor can influence the establishment of different soil organisms, change the effects of topography, and even create subtle differences in the atmosphere that may affect climate. Turf affects its environment too. Turf plants complete the cycles of nutrients, water, and soil life, providing environmental stability. Most environments are still evolving, but plants act like a sea anchor, stabilizing and often removing much of the volatility from the evolutionary process.

Soils are subject to eventual ruination by the same forces that created them. The inherent mineral content of any soil may be vast, but it is also finite and can eventually be depleted by weathering. At some point, perhaps a billion years from the time of a soil's creation, leaching, erosion, or both can eventually deplete the mineral resources that are crucial to the existence of living organisms. As the availability of minerals and organic residues wanes, so does the biomass that sustains the level of native organic matter which structurally protects the soil from erosion and stores much of the moisture and nutrients necessary for biotic development. As depletion continues, only the most weather-resistant particles of quartz and other mineral compounds remain. A rich soil can eventually evolve into a habitat that sustains only the hardiest of organisms.

Erosion, one of the forces that weathered rock and carried particles of parent material to deposits called soil, can also carry soil away. Often soil can be carried all the way into an ocean, where tremendous pressures for long periods of time can transform the mineral particles back into sedimentary rock. Perhaps, a million years from now, a broad tectonic event will lift the recreated rock into a mountain and the process can begin all over again. A crucial component of the soil ecosystem that retards weathering and erosion is plant life and the diverse population of soil organisms it supports. Glue-like substances created by many soil organisms aggregate soil particles and protect them from erosion and other weathering forces.

Plants also protect the soil environment in a number of different ways. They provide shade and slow the evaporation of precious moisture needed for their own lives and the lives of other soil organisms. The shade they provide also reduces soil temperature which can increase net primary production (i.e., the production of energy beyond the plant's own needs). The extra energy produced feeds soil organisms that also participate in soil conservation. Plants also control erosion with their root systems and the organic matter they produce. The adhesive qualities of humus hold soil particles together and provide greater absorption of water through an increase in soil porosity. This increase in water absorption provides a relative decrease in surface runoff.

The mechanism by which plants produce the proteins, sugars, starches, fat, and fiber needed by the consuming species of the world is chemically quite complex. A plant's ability to use the endless energy from the sun and the small percentage of carbon dioxide in the earth's atmosphere to produce all of these organic nutrients is nothing short of a miracle. In simple terms, small openings called stomates, mostly on the underside of the plant's leaves, allow the transference of gas and moisture into and out of the plant. Plant leaves intercept carbon dioxide, much of it generated by the biological decay of organic residues, as it escapes the soil. Plant cells that contain chlorophyll are excited by the energy of the sun, and conduct chemical reactions within the leaves. These reactions combine the carbon and some of the oxygen from carbon dioxide (CO_2) with hydrogen from water, nitrogen, phosphorus, sulfur and other minerals from the soil, to form the complex organic compounds that give life and growth to the plant, to its roots, and to every other organism that depends either directly or indirectly on plants for sustenance. The six elements mentioned above—carbon, oxygen, hydrogen, nitrogen, phosphorus, and sulfur—make up 99% of all living matter.

Other nutrients considered essential to plants are: calcium, molybdenum, magnesium, boron, copper, iron, zinc, chlorine, potassium, and manganese. It is important to note that elements not considered essential for plant tissue development may be of vital importance to other soil dwelling organisms, which play their own crucial roles in the dynamics of the plant growing system (see Chapter 2: Fertility).

All of the carbon that plants use is derived from the atmosphere. Oxygen comes to the plant from the atmosphere (as carbon dioxide) and from water (H_2O), which also delivers hydrogen. These three elements—carbon, hydrogen, and oxygen—constitute the basic building blocks of all organic compounds and comprise approximately 95% of a plant's diet. The remaining 5% is in mineral form and derived from the soil (see Figure 1.11).

FIGURE 1.11. PLANT NUTRIENT NEEDS

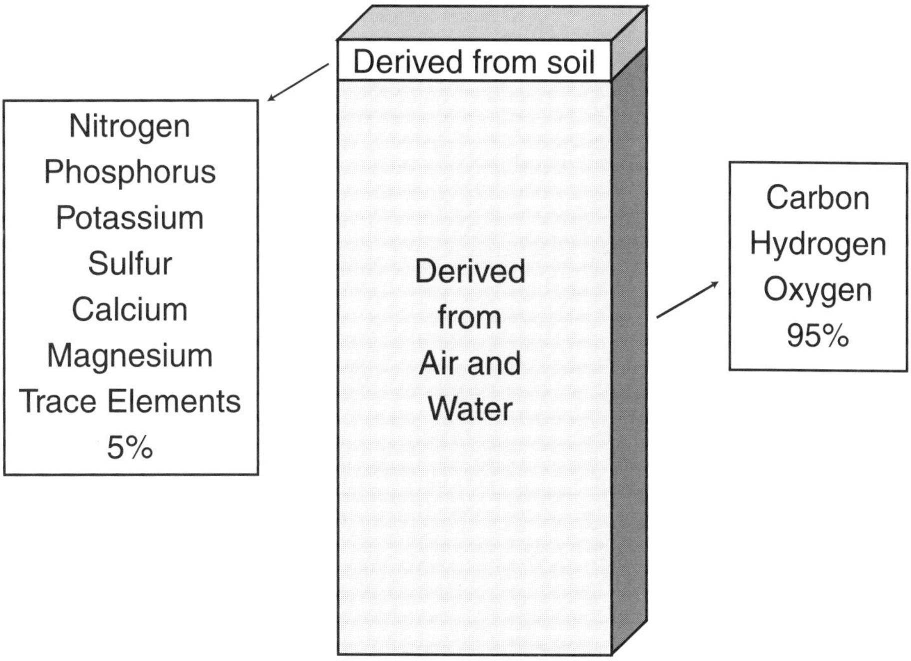

Derived from soil

Nitrogen
Phosphorus
Potassium
Sulfur
Calcium
Magnesium
Trace Elements
5%

Derived from Air and Water

Carbon
Hydrogen
Oxygen
95%

Weathering forces, soil organisms, climate, plants, and animals (including humans) have had, and continue to have, a profound effect, both direct and indirect, on the soil's evolution. The existence of almost all life is both controlled by, and controls, the biological, physical, and chemical diversity in the soil. The superintendent certainly belongs in this equation so it is important that he has a basic understanding of the soil ecosystem. The overview presented here is painted with a broad brush. Volumes of information—some of it conflicting—have been published about the soil and only a brief scan of that material could be included in this chapter. But all the currently available information on soil probably represents only a fraction of what has yet to be discovered. The soil is indeed a complex ecosystem but its biological health and vitality appear to be inextricably linked to the vigor and resilience of turf. Recognizing the living component of the soil system as the most ideal barometer of soil health may be the extent of what we need to understand. It's possible that all we have to learn now are the best ways to care for it.

Chapter 2:

Fertility

Before discussing the appropriate diet for a full-grown, thoroughbred turf, it would be worth examining how plants managed to survive before humans intervened. There's no argument that turf needs fertilizer but the nitrogen, phosphorus, potassium, calcium, magnesium, sulfur, and all the trace elements that superintendents apply, surprisingly, only account for 5% of the plant's total diet. The other 95% comes from the atmosphere and from water. As relatively minor as their mineral requirements may seem, the question remains: How does the plant acquire the nutrients it needs if no one is there to apply it? Obviously, plants have been surviving on their own for millions of years, well before humans or any other beings with fertilizer spreaders appeared on earth.

Carbon and oxygen from carbon dioxide in the air and hydrogen from water combine to make the building blocks of all the organic compounds produced by photosynthesis. All of the other necessary nutrients are digested from organic and mineral sources in the soil and made available to plants by many different groups of soil organisms. Proteins, carbohydrates, fats, waxes, cellulose, and other organic compounds found in plant and animal residues, in soil organic matter, and in the soil's biomass, and minerals contained in inorganic soil particles, all contribute essential nutrients to plants. But soil organisms must first digest and transform these nutrients into an available form. These organisms in the soil function as the plants' digestive systems. Much of the carbon dioxide in the atmosphere is generated in the soil from biological activity. Just as humans and animals release carbon dioxide as a respiratory waste, so do soil organisms, and the grass plants' best opportunity to absorb carbon dioxide is when it first emerges from the soil's surface before atmospheric turbulence can dilute it with other gases. This delivery system has, over the millennia, proved to be effective, efficient, and difficult to improve upon.

Unfortunately, this system is not functioning at or even near peak performance on most areas of most golf courses. Typically, there is not a lot of organic matter in the golf course's soil, which limits the population of soil organisms and the creation of biomass, and not very much in the way of organic residues is ever applied to the soil. Even grass clippings are often removed from greens, tees, and sometimes from fairways as well. Soil organisms have very limited resources for themselves, which limits their ability to make nutrients available to plants, and the production of carbon dioxide is relative to their level of activity. Realistically, when we discuss fertility we should consider the needs of the entire plant growing system, which includes soil organisms.

Normally, fertilizers are chosen based on the N-P-K analysis; that is, the percentage, by weight, of total nitrogen, available phosphate, and soluble potash. Superintendents may also apply secondary macronutrients—calcium, magnesium, and/or sulfur, and trace elements if a soil test or leaf tissue analysis indicates a deficiency. It is a rare superintendent indeed who looks for a fertilizer that provides digestible carbon for soil organisms and even more unusual is a superintendent who wants to know whether the carbonaceous fertilizer will favor populations of bacteria or fungi. These are, however, important questions to ask. The right balance of organisms living in the soil provide turf plants with both sustenance and protection, and, as carbonaceous residues are digested, these organisms also generate carbon dioxide that grass plants can capture as it permeates through the soil's surface. Adequate populations of these groups of organisms cannot be sustained unless the proper resources are available.

Weed, insect, and disease problems or poor turfgrass performance are often symptoms of an incomplete or deficient soil system. Pests are typically kept in check with pesticides and poor turfgrass performance is normally dealt with by adding chemical nutrients. But the problem causing these symptoms may very well be biological. Biological imbalances are generally less problematic on a golf course than the more typical status of completely inadequate levels of biological activity.

Carbon in the soil is found in the living biomass, organic residues, friable humus, and stable humus. These sources provide energy for most soil organisms. Soil carbon is produced by autotrophic organisms such as plants and algae that can combine carbon from the atmosphere with energy from the sun and produce organic stuff. The carbon compounds produced by autotrophs eventually become part of a vast warehouse of energy and protein known as soil organic matter. This warehouse functions beneficially in hundreds of different ways, but two essential purposes are to provide energy for soil life and carbon dioxide for plants.

When fresh organic residues come in contact with the soil, decay begins almost immediately—at least during the seasons that soil organisms are active. What determines the speed at which organic residues are decomposed is temperature, moisture, air, and the balance of carbon and nitrogen in the residues. The carbon to nitrogen ratio (C:N) is always measured as x parts carbon to one part nitrogen. If the C:N ratio is wide (i.e., high carbon), as it is with straw or wood chips, decomposition occurs slowly. Furthermore, soil organisms need to commandeer available nitrogen from the soil as they digest the carbon because their diets need a narrower C:N range. This usurpation temporarily overrides the nitrogen needs of plants. If the original organic litter has a narrow C:N ratio, as it does with grass clippings, food wastes, or animal wastes, decomposition will likely occur more rapidly, making nitrogen available to plants and other organisms. Each time saprophytes (decay organisms) digest components of organic residues or organisms are consumed by predators, some energy is used and carbon is oxidized into carbon dioxide (CO_2), which is then released into the soil atmosphere.

Turf lives healthiest in a biologically balanced soil—one in which bacteria accounts for slightly more than half of the living biomass. Bacteria thrive in soils that contain residues high in carbohydrates, that is, sugars and starches. Vegetable and animal meals and compost made from food waste, manure, or biosolids contain the carbohydrates necessary to sustain a bacteria dominated soil. Yard waste contains more cellulose and lignin, which can favor fungal domination. Yard waste compost is a good soil conditioner but may not be the most ideal for a golf course where a more even balance between bacteria and fungi is preferred. According to Cornell University research, yard waste compost does not offer as much disease suppression as composts made with less woody ingredients. However, there are procedures that can change the balance of bacteria and fungi in a yard-waste compost pile (see Chapter 3: Compost).

Feed the soil. We have all heard this axiom before—sometimes from people who have an organic garden, a wood stove, and a composting toilet. Whether you believe them to be credible or not, their philosophy is sound and it underlines the basic needs of a functioning soil ecosystem. We usually pick fertilizers based on the guaranteed analysis, which tells us the total nitrogen, available phosphate, and soluble potash. Sometimes magnesium or some trace minerals like iron or manganese are added. If we are to feed the soil, we need to apply food such as carbohydrates and proteins—ingredients that are impossible to find on fertilizer labels. The essential nutrients we apply are important but without a healthy, functioning ecosystem, many of these resources may be utilized inefficiently.

Nitrogen (N) is a great example of this. Plants use nitrogen for many different functions. Without N, plants could not manufacture peptides, amino acids, proteins, enzymes, chlorophyll, or nucleic acids (the building blocks of DNA and RNA). These components are essential for plants to grow and function properly. The reactions that occur during photosynthesis require a lot of nitrogen. If N is deficient during photosynthesis, less chlorophyll is produced and plants begin to lose their green color (chlorosis). Turf responds to many stress conditions by producing special proteins and other metabolites that are rich in nitrogen. Inadequate access to N renders plants more susceptible to stress-related problems. Plants also produce special enzymes to combat pathogen infection and even insect attack. Enzymes, like protein, need N. On the other hand, excessive applications of nitrogen can also contribute to problems. Plants respond to abundant N by increasing top growth and arresting root growth. Free amino acids often accumulate in the leaves and invite both foraging insects and disease pathogens. In addition, a guttation fluid that is rich in nitrogen and other nutrients can exude from the leaf tips—a phenomenon that is analogous to raising a banner that reads, "EAT AT JOE'S." When the guttation fluid dries, the leftover salts can burn the leaf tips. Another potential for problems is that when plant growth is pushed with excess nitrogen, the production of their defense compounds is often suppressed.

Applying just the right amount of nitrogen is not only extremely important but it is almost impossible because conditions that determine the amount of N a plant needs at a given time are constantly varying. Air and soil temperature, moisture, HOC, the angle and intensity of the sun, biological production of carbon dioxide, and the variety of grass all affect the rate at which turf plants use nitrogen. The optimal amount of N today could be excessive or inadequate tomorrow. Plants have the ability to respond to both deficient and excess nitrogen but it is not always enough. When nitrogen is inadequate, plants produce sugars in the leaves that are quickly transported into the roots. These sugars feed roots and enable them to grow farther into the soil in search of more nitrogen. If too much N is available, the leaves produce amino acids which increase shoot growth. Sugar production is all but arrested and root growth is suppressed. This slows down the absorption of N through the roots.

Plants don't understand the difference between nitrogen that is biologically released from organic sources and the soluble kind that is applied by modern superintendents. Nitrogen is nitrogen to the plant. But they also have a hard time understanding and adapting to the feast to famine scenario associated with many chemical-feeding programs. During periods when N is inadequate plants respond by elongating roots. When a tidal wave of nitrogen becomes available from an application of soluble N, and an extra-diffusive root system absorbs more than the plant needs, some serious side-effects that include disease susceptibility, insect attraction, burning, and other problems can occur. In a healthy soil ecosystem, however, there are mechanisms that buffer and regulate the amount of N available to plants.

Many soil organisms need nitrogen more than plants but they need it balanced with a certain amount of carbon. This balance occurs in proteins that reside in organic residues, friable humus, and in the prey upon which many soil organisms feed. In fact, one of the main ways in which organisms make N available to plants is through predation of other organisms. Predators such as bacteria-feeding nematodes, for example, don't need or want the amount of nitrogen inherent in their prey and release the excess as ammonium, which is then biologically converted into nitrate that turf roots can absorb. Applications of compost or natural organic fertilizers complement the functions of soil organisms and they, in turn, regulate the amount of N available to plants. This is not to say that soluble N should never be used but if it is the only source of nitrogen applied, biological activity will likely be suppressed to a point where nutrient regulation may be inadequate. There has to be available digestible carbon if soil organisms are going to regulate applied nitrogen. If the turf ecosystem is healthy and functioning properly, judicious applications of soluble N can be regulated by soil organisms. Soluble nitrogen can be used by many soil organisms which, in turn, moderates the amount available to plants. As predation and other biological activities occur, more nitrogen is released for turf roots to absorb. Coincidentally, the activity level of the organisms that release N and other available nutrients respond to many of the same conditions that stimulate plant activity so the availability of nutrients is often synchronized with need. Spoon-feeding soluble N is rarely as ideal.

If large doses of soluble N are applied, both plants and organisms can be overwhelmed. High salt fertilizers can act like a dry sponge, drawing moisture away from the surrounding area. This can cause osmotic shock, killing organisms in close proximity. Eventually, as the salt is dissipated within a greater solution, the dissolved nutrients change from harmful to helpful and can be assimilated by some soil organisms. Once absorbed into a living organism, the nutrient is somewhat stable until the organism dies or is preyed upon. If it dies, nitrogen may or may not be released depending on the saprophyte that consumes the remains. If the saprophyte requires less N than what is available from its food source, then some will be released; however, if it requires more, then nitrogen may be immobilized from other sources. In a turf soil, the former scenario is more likely than the latter. If a predator eats the organism, there is almost always a release of N in a form available to plants. Not only can this system increase nitrogen efficiency but it can also sustain the release of it for a longer period of time. Unfortunately, very little of this biological regulation can occur on a green constructed of sand, topdressed with sand, and fertilized purely with soluble salts. There needs to be a healthy, active, well-fed, and diverse population of soil organisms and this is nearly impossible to achieve or maintain in an environment without adequate resources. The sand-based environment on many greens serves as an almost inert medium that provides mechanical support for the turf plants. It is rarely cultivated as a habitat for soil life.

Developing a program that addresses the biological needs of the soil is a logical step toward improving fertility. The use of compost in the topdress mixture, compost tea in the irrigation system, and natural organic fertilizers in the spreader can not only reduce the need for soluble N but increase its efficiency as well. In other words, less soluble nitrogen can accomplish more. Applications of compost during spring and fall core cultivation incorporates valuable resources for a diversity of soil organisms. The addition of compost tea and natural organic fertilizers provides more nutrients and even some inoculation. If soluble N is still needed, small doses can react synergistically with biological functions that increase overall nitrogen efficiency. It may not be needed, however. Well-made, mature compost can contain anywhere from ½ to 3% nitrogen depending on the feed stock from which it was made and the moisture content. Nitrogen content in compost is usually measured on a dry basis and the value found is a percentage of the material's weight, so if a dry compost contains 1% N but its as-is moisture is 50% (of its weight), then the actual N content is 0.5%. If a topdress layer consisting of 50% sand and 50% compost is applied at a thickness of ⅛ inch, then almost 0.4 yd^3 of material (~0.2 yd^3 sand and ~0.2 yd^3 compost) has been applied per 1,000 ft^2. If the weight of the compost per yd^3 is 1,500 pounds and it contains 0.5% of nitrogen (on an as-is basis), then almost 1½ pounds of N have been applied per 1,000 ft^2. Don't expect a great flush of growth, however. The release of N from compost is slow and sustained and depends on factors such as temperature and moisture. If turf is growing in a very cold soil, some soluble N may be necessary to fulfill the plants' needs until the soil warms up.

Natural organic nitrogen is synonymous with protein. The main reason that it is more expensive than its chemical cousin is that it has a greater value in animal feed and pet food markets than it does as fertilizer. Protein is not an available nutrient for plants. The chemical structure of protein is too large and complex to be assimilated directly through plant roots. Protein, however, can be assimilated by soil organisms. When proteinaceous materials come in contact with the soil, organisms begin dismantling it into amino acids and peptides. Most of the nitrogen and carbon is consumed and temporarily immobilized but some is released and mineralized by other organisms into simple nitrogen ions that plant roots can absorb. The population of saprophytes that consume the protein grows exponentially which results in a relative increase in the number of predator organisms that feed on them. The result of this increased predation is an increase in available nitrogen for plant roots but the release is steady and sustained unlike the tsunami of pabulum that is typical of soluble fertilizers. An additional advantage to using proteinaceous nitrogen is that it is extremely efficient. Very little if any is ever lost to leaching, volatilization, or denitrification and the corresponding increase in biological activity can often reduce plant susceptibility to disease infection and attractiveness to some foraging insects. Taking into consideration its efficiency and its ability to nurture an active soil food web (which can lead to significant savings elsewhere), the value of natural organic nitrogen can be considered more relative to its price.

Researchers have found that soil where unfertilized turf is growing contains more soluble nitrates during the late summer. It would not be unreasonable to assume that applications of soluble N at this time of the season are more prone to leaching. Natural organic nitrogen may be a much more appropriate material to use at this time of year.

Natural sources of N include blood meal, feather meal, meat and bone meal, hoof and horn meal, whey solids, leather meal, compost, cocoa meal, soybean meal, cotton seed meal, manure, biosolids, crab meal, fish meal, grass clippings, castor pomace, coffee grounds, alfalfa meal, and peanut meal. In fact, most organic residues from plants or animals contain some nitrogen but sometimes not enough to be valuable as a nitrogen source. Manure is not an amendment we would recommend applying anywhere on a golf course except on the compost pile. It would be an unusual raw manure that doesn't contain a cornucopia of weed seeds.

Constant and excessive infusions of soluble nitrate or ammoniated fertilizers cannot only do more harm than good in the soil but can cause pollution in shallow groundwater as well. Moreover, a significant amount of money may be wasted on the percentage of these materials that never reaches the plant. Consequences of excess nitrogen for turf include higher susceptibility to disease and insect pests; excessive thatch development; decreased tolerance to environmental stress conditions such as heat, drought, and cold; poor root and lateral shoot development; and reduced production and reserves of plant carbohydrates.

In addition to improving the efficiency at which plants use nitrogen, a diverse and prosperous soil food web can also bring some free nitrogen down from heaven. Free living, nitrogen-fixing bacteria—those not associated with the roots of legumes—can consume atmospheric N for the production of proteins, enzymes, etc. When they die, or are consumed by predators, that nitrogen is cycled through the soil ecosystem and eventually becomes available to plants. Measurements taken back in the '60s show fixation of between 20 and 100 pounds of nitrogen per acre per year by nonsymbiotic bacteria. The research also found a corresponding relationship between the amount of organic residues returned to the soil and the amount of N gained from free-living, nitrogen-fixing bacteria. The more residues made available to soil organisms, the more nitrogen fixation occurred. Interestingly, the study found that the highest level of N-fixation occurred in a bluegrass sod. There were no details in the reference (Brady, 1974) about how the sod was cared for or its ultimate use but the findings suggest that there is great potential for turf to feed a prolific web of soil organisms.

Even though golf course conditions aren't found anywhere in nature, contemporary fertilization practices do not in any way resemble natural feeding phenomena. The use of soluble, available nutrients bypasses the digestive system of plants, i.e., the living biomass in the soil. A fertility program that ignores the needs of these organisms is, in a sense, repudiating their value and importance. The assumption that these soil organisms have little to no worth may be the main reason why managers in both agriculture and horticulture have become dependent on chemicals.

The availability of most other plant nutrients also has direct and indirect relationships with biological activity in the soil. Phosphorus, for example, is a nutrient that plant roots can have a hard time finding without the help of soil organisms. Most turfgrass experts will tell you that established grass does not need a lot of phosphorus because P is needed mainly for germination, flowering, seed production, and initial root development—unnecessary or unwanted activities for a mature stand of turf. In fact, applications of P in excess of what turf needs may only serve to increase the germination of annual weed seeds. Phosphorus is, however, an extremely important element for turf plants and all other living organisms. Adenosine triphosphate (ATP) is a component that is involved in reactions that transfer energy throughout the plant. The genetic information components, DNA and RNA, also contain phosphorus. Phospholipids form the basic structure of cell membranes and play a role in the selective movement of nutrients and other materials into and out of the cell.

The activities of soil organisms make phosphorus available to plant roots in many different ways. Phosphorus that is tightly fixed with other elements can be released when the acids, enzymes, or chelates produced by biological activity make contact. Humates produced by soil organisms can prevent P from fixing with other elements by complexing with phosphate ions or by interrupting the sites where fixation might occur. The production of carbonic acid that results from biomass respiration can also free fixed phosphate ions. The introduction of fresh residues often stimulates the decomposition of some humus containing phosphorus. Additionally, the abundance of root hairs that are typical in biologically rich soils have significantly greater opportunity to encounter phosphate ions. The likelihood of these scenarios occurring, however, is relative to the amount of resources the soil holds for members of the food web. In a biologically inactive soil, it is often necessary to apply P in higher quantities than what is actually needed to compensate for what will be immediately fixed to other soil ions. Some of the ions such as iron, calcium, or magnesium that bond with P are also rendered unavailable to turf roots.

Most of the phosphorus used in fertilizers is refined from phosphate rock (PR). PR is a mined marine deposit that typically contains ~30% total phosphate but only 3–8% is considered available—depending on the method used for analysis. The amount of available phosphate from PR is generally higher when it is introduced into a biologically active soil. Eventually, over a period of three to ten years, most, if not all, of the phosphate contained in PR will likely be released. The advantage to this very slow, long-term release is that it need not be reapplied more often that every half decade (if at all). Conventional fertilizers contain refined derivatives of PR that dissolve easily and move quickly into the soil solution. The availability of phosphate from these materials is much higher and, consequently, what plants can't use can quickly fix with other elements or compounds in the soil and become unavailable. Soil reactions that release phosphate from PR do so slowly and over time plant roots receive small but consistent doses of this essential element. If needed, PR is usually fine enough for greens and can be applied alone or as a component of a topdress mixture.

When phosphate ore is first mined, it is washed to remove some impurities (mostly clay). The effluent from this operation is deposited in lagoons where its moisture evaporates in the sun. The remaining material is almost as rich in phosphate—two to five % available, 18% total—as the washed PR, but also contains colloidal clay. The clay is called colloidal because it can bind coarse soil particles together and increase the soil's CEC. It is relatively rare that a superintendent would use colloidal rock phosphate (a.k.a. soft rock phosphate), especially on greens, but a few managers put a small amount in their topdress mixture to conserve water use and increase the CEC.

Compost usually contains a small amount of phosphorus (<1%) but, again, the volume that is usually applied more than makes up for the low analysis. The P in compost is organically bound and biologically released. The activity that makes P from compost available to plants is normally synchronized with plant need so efficiency is increased.

Most soils contain adequate levels of phosphate for plants but very little of it may be available if soil organisms are relatively inactive. Research has shown that the availability of P—along with many other elements—increases when organic residues (food for soil organisms) are introduced into the soil and that the increase exceeds the amount of P contained in the residues. Additionally, mycorrhizal relationships are often established with roots growing in a biologically active soil. Mycorrhizae are great retrievers of P (and other nutrients) from the soil. Not only do their hyphae explore far greater regions of the soil than roots do but they secrete powerful acids and enzymes that solubilize P from fixed locations. The beauty of this system is that mycorrhizae can only deliver phosphorus to the roots to which they are attached. They do not supply P to germinating weed seeds and do not normally colonize annual weed roots. The organic fraction of the soil, which includes stable and friable humus, plant and animal residues, and the living biomass, contains a significant amount of P in a form that is more easily mineralized than most of the tightly-bound inorganic sources. Soil organisms continually cycle this phosphorus through the soil ecosystem . . . that is, if they are active enough. Earthworms increase P availability by 500–1,000%, and by increasing the soil region to which turf roots have access (through earthworm tunnels), plant roots have greater opportunities to find P-containing sites.

Like phosphorus, potassium (K) is also regulated to a large degree by soil organisms. Potassium is an extremely important nutrient for turf, especially if the plants have to endure the rigors of golf—both the game and the maintenance. Potassium gives plants the strength to stand upright and better endure both natural and man-made stress. Disease resistance is often credited to adequate levels of K. Unlike phosphorus and nitrogen, potassium doesn't combine with other elements or compounds in the plant to form proteins, carbohydrates, or other products of photosynthesis. It lives by itself in the plant fluids and is responsible for maintaining the correct hydraulic pressure within plant cells and assists with the accumulation and translocation of photosynthesized materials. Without adequate potassium, nitrogen cannot be used efficiently by the plant. In fact, some of the detrimental effects of excessive N are regulated by potassium.

Unfortunately, soil conditions on most greens are not ideal for holding a good supply of K. Available potassium usually resides in the soil solution and on colloidal exchange sites (see Chapter 1, Soil Ecosystem). Colloidal exchange sites are (negatively) charged particles of certain types of clay or stable humus, neither of which are abundant in sand greens. Potassium in soil solution is susceptible to leaching out of the root zone especially in soils designed to drain rapidly. Liming can help slow the loss from leaching but can also reduce the amount of K available to plant roots. Potassium is also contained in the cells of soil organisms and their life cycles make some available to plants. There is a tremendous amount of naturally occurring potassium in most mineral soils but the vast majority is unavailable to plants. Some of this bound potassium can be released through the activities of soil organisms. Secretions of their acids and chelates can release K from insoluble mineral sources. These biologically freed potassium ions are generally assimilated into the biomass but some inevitably become available to plants. There is also abundant potassium fixed between layers of certain types of clay. These ions are less tightly bound than the insoluble mineral sources but are much less available than K ions attached to soil colloids or in soil solution.

In a sand green there is usually no clay, and very little soil or humus. Even if there were an abundance of fixed potassium in the soil, biological activity would probably be inadequate to release substantive amounts of it. Soil organisms do more than just release fixed sources of K; they create humus that can magnetically hold potassium ions in an available form and encourage plants to grow finer, more abundant root hairs. Finer roots have greater surface area and can absorb potassium and other nutrients more efficiently. The production of stable soil humus can only be accomplished by soil organisms. It is not a material that is commercially available and, if it were, moving and handling it would destroy most of its beneficial properties. There are many products on the market containing clinoptilolite (see Chapter 6, Cultural Practices) that claim to raise the cation exchange capacity and increase the soil's capacity to hold cation nutrients like potassium. Research is generally positive but these products are relatively expensive to use and there is evidence that a few of them can inhibit the diffusion of water through the soil. Humus is the quintessential colloidal particle. Adequate levels of humus in the soil do not inhibit drainage beyond an acceptable threshold but offer tremendous benefits to the soil ecosystem. The only way to introduce humus to the soil is to add organic residues to the biological, humus-manufacturing machine on a regular basis.

As the pH of a soil increases so does the capacity of humus particles to hold cations. At optimum pH for turf, the CEC of humus is often at least twice that of clay particles.

A near neutral pH will normally reduce the amount of potassium in the soil solution and reduce the potential for leaching but it can also reduce the amount of K that is available to turf roots. A soil that is rich in humus, however, encourages diffusive and abundant rooting accompanied by a prolific biomass congregating in the rhizosphere. The greater root surface area gives turf plants better access to potassium that is in the soil solution and attached to humus particles, and the biological activity in the area immediately surrounding the roots lowers the pH in the rhizosphere and increases the availability of K. This natural system is difficult to improve upon.

Potassium moves by diffusion; that is, it moves from areas of high to areas of low concentration. If the amount of K in the soil solution wanes, potassium ions move from the exchange sites into soil solution or vice versa and it does this very quickly. If, however, the CEC of a soil is extremely low (typical in sand greens) there are inadequate storage sites and movement of K is normally from the fertilizer spreader to soil solution. From there, roots absorb some and the rest moves away from the root zone as the soil solution percolates downward. Capillary activity can move some soil solution back up to the root zone but constant irrigation disrupts upward mass flow. Adequate moisture in the soil is also crucial for diffusion to occur but excessive water decreases utilization if oxygen levels fall below a certain threshold. As oxygen levels in the soil decrease, roots become stressed and nutrient uptake is reduced but the absorption of potassium is effected more than any other nutrient.

Plants are like most dogs when it comes to potassium in that they will consume far more than they need. This presents an efficiency problem for turf growing on sand greens. Consumption beyond optimum doesn't improve quality but the turf plants have little choice. Since colloidal particles are scarce in a sand green, applied soluble K can only reside in the soil solution and because of the superfluous drainage through sand, plants are faced with a use-it-or-lose-it situation. Turf plants do not contemplate their choices, of course, but they seem unable to regulate the amount of potassium they consume. The excess K that roots absorb is not stored and used when needed. It is essentially wasted. A higher CEC from the presence of stable humus and a greater population of soil organisms would better regulate the availability of K and plants would have more efficient access.

The most common sources of potassium (potash) are mineral salts such as potassium chloride, potassium sulfate, or magnesium/potassium sulfate, all of which are considered soluble. Once dissolved, the potassium enters the soil solution and is absorbed by turf roots, adsorbed onto exchange sites (if there are any), or remains in solution and eventually leaches out of the root zone. Insoluble sources of potassium include greensand and other rock dusts but these materials are slow to break down and it is unlikely that the immediate K requirements of turf plants will be met, especially in an environment with minimal biological activity. Some managers add these rock powders or ashes of wood or seed hulls to their compost pile where biological activity is abundant. Compost typically contains 0.2 to 1.0% potassium on a dry-weight basis, which doesn't seem like a lot but given the average compost application, can add up. A 3⁄16 (0.1875) inch topdress application that consists of 50% compost may delivery a pound or more of potash per 1,000 ft^2. The potassium in compost is located in the biomass, those living organisms that inhabit compost, and is chelated to organic molecules of protein, fatty acids, lipids and others. These ions are released biologically. Not only is this delivery system more efficient than applications of soluble salts but the compost can eventually contribute to the CEC of the soil and further increase the efficiency of applied potassium salts over time.

Calcium (Ca) and magnesium (Mg) are cations like potassium and reside in soil solution and on colloidal exchange sites. Both Ca and Mg move through the soil by diffusion like K but can also move with the flow of water through the soil. Balancing Ca, Mg, and K in the soil is also important because they compete for space on exchange sites; the abundance of one can cause the deficiency of another (see Chapter 4, Analysis). Potassium, the biggest ion, can displace calcium or magnesium but excess K will likely result in a Mg deficiency first. An ideal ratio of Ca, Mg, and K for turf is ~80:11:3, respectively, but this is not to say that one cannot have a healthy stand of grass at a slightly different ratio. Since magnesium is the weakest of the three cations—in terms of exchange site competition—many believe that an overage of Mg in the soil will not create any problems. Anecdotal evidence, however, indicates the important of maintaining the correct ratio, especially between Ca and Mg. Many agronomists believe that a Ca:Mg ratio that is narrower than 7:1 (but not wider than 10:1) can result in an eclectic but indirectly related group of symptoms that include compaction, disease susceptibility, insect attraction, and stress intolerance.

A narrow Ca:Mg ratio (<7:1) is often caused by overuse of dolomitic lime. Most plants consume significantly more calcium than they do magnesium and reduce the soil's inventory of calcium at a faster rate. Dolomitic lime has a Ca:Mg ratio of ~2:1 which, over time, can cause a buildup of magnesium. The opposite (too wide a Ca:Mg ratio) can occur from high calcareous sand content in the soil. Calcium buildup from this type of sand can create shortages of Mg, K, or both. In this situation, the use of either calcitic or dolomitic lime would be unwarranted and applications of potassium/magnesium sulfate, potassium sulfate, or magnesium sulfate may be necessary to improve the Ca:Mg:K ratio.

Calcium helps give plants structural rigidity which is extremely important for recreational turf. It also regulates root and shoot growth and the development of several important enzymes. Magnesium functions as a crucial component of chlorophyll, activates many essential enzymes, plays a role in protein synthesis, and in the production of ATP. Consistent availability of Ca and Mg is crucial to the health of turf plants but as with potassium, the storage facility in a sand green is less than ideal.

Unlike potassium, calcium and magnesium can be supplied by liming which naturally releases ions into the soil solution at a relatively slow and sustained rate. This release rate would be more effective and efficient in a biologically active soil environment. The system of exchange that nature created protects plants from deficiencies of vital cation nutrients. Without adequate storage capabilities in the soil, periodic shortages of cations can occur which may trigger symptoms that are not necessarily indicative of a nutrient deficiency. If, for example, a short-term deficiency of calcium suppressed root development enough to stress the plant but not enough to show symptoms of a calcium deficiency, the result may be disease susceptibility, drought intolerance, or just general decline. Having a constant and adequate supply of these nutrients available to turf plants is important and difficult to accomplish without sufficient exchange capacity in the soil. Regular applications of compost in the topdress mixture can slowly build an adequate exchange mechanism in the soil without jeopardizing drainage.

Sources of calcium include gypsum (calcium sulfate), phosphate rock (tricalcium phosphate), super- or triple-super phosphate (dicalcium phosphate), calcium nitrate, calcium oxide, and lime (calcium carbonate). Dolomitic lime (calcium/magnesium carbonate) also contains magnesium as do Epsom salts (magnesium sulfate), potassium/magnesium sulfate, and magnesium oxide. There are occasions when calcium or magnesium may be needed but raising the pH is not necessary or desirable. Gypsum, Epsom salts, and potassium/magnesium sulfate can add Ca and Mg cations without altering pH. The use of any of the three also adds sulfur to the soil which, in some areas, may be deficient. Calcium phosphates, calcium nitrate, calcium oxide, and magnesium oxide can alter the soil pH but application rates designed to fertilize are usually not large enough to significantly alter soil pH. The use of phosphates or nitrates where only calcium is needed may introduce excessive phosphorus or nitrogen.

Sulfur (S) is a plant nutrient that doesn't receive as much attention as it should. Sulfur is available to plants from many different sources. Organic matter contains sulfur, as does the mineral component of some soils but surprisingly, most of the available S has, in recent past, come from air pollution. Sulfur dioxide in the atmosphere is a by-product of burning fossil fuels, especially coal, and can settle to earth in significant quantities. The deposition of S from air pollution is usually greater closer to the source of pollution but ranges from less than four pounds per acre, in more rural areas, to as much as 175 pounds per acre closer to industrial sites. Sulfur has also been used in the formulation of many pesticides and is a component of some fertilizers such as potassium-, ammonium-, calcium-, or magnesium sulfate. Deficiencies of sulfur have been relatively rare and because of their infrequency, the symptoms are often difficult to recognize. In recent years, however, the release of sulfur dioxide into the atmosphere has been abated and the use of sulfate fertilizers and sulfurous pesticides has also decreased significantly. Add to those phenomena the aversion many superintendents typically have toward organic matter in greens soil, and the possibility of sulfur deficiencies for turf plants significantly increases.

Sulfur is a component of many important amino acids without which some vital proteins cannot form. It also is a constituent of some essential vitamins and important enzymes that mediate crucial plant transformations. Some of these enzymes play a major role in photosynthesis. Sulfur deficiencies can result in suppressed photosynthesis and show signs of chlorosis much like symptoms of a nitrogen deficiency. Without sulfur plants cannot produce chlorophyll, lignin, pectin, and other important plant components.

Most of the sulfur in soil resides in soil organic matter. Mineral soils rich in pyrite also contain sulfur reserves but neither pyrite nor organic matter is found in ample quantities on a sand green. Like nitrogen, sulfur is largely stored and regulated by biological activity. Under some circumstances, soil organisms release sulfur for plants and under others they immobilize it. This phenomenon is usually controlled by fluctuating ratios of carbon to sulfur (C:S) in the soil. When the C:S ratio is greater than 400:1, organisms generally immobilize more sulfur than they make available to plants. As the ratio decreases from 400:1 to 200:1, there is a relative increase in the release of S and a decrease of S immobilization. Below 200:1, organisms release significantly more sulfur in a plant-usable form than they immobilize. The ratio of C:S in soil organic matter varies but averages ~250:1. The variation of C:S ratio in organic residues is also wide and that range is relative in compost but analysis of compost samples reveals an average C:S ratio of ~200:1 which would not only provide for the release of S from itself but narrow the overall ratio in the soil (depending on how much and how often it is applied) for more sustained release of necessary sulfur. The adequate availability of S for plants depends on several factors including temperature, moisture, carbon/sulfur ratio, and pH but the most influential factor is biological activity, which is stimulated by regular contributions of organic residues to the soil. It is not possible to calculate the C:S ratio from the data on a soil analysis. Digestible carbon would never appear on a standard soil analysis and if sulfur were measured, it would be as available—not total.

Plants absorb sulfur as sulfate (SO_4). Like potassium, sulfate moves through the soil by diffusion and, like magnesium, and calcium, it also moves by mass flow (i.e., the movement of water through the soil). Unlike potassium, magnesium, and calcium, sulfate is not a cation—a positively charged ion—which can cling to colloidal exchange sites. Sulfate is therefore at a much greater risk of leaching than potassium, magnesium, or calcium. In a sand green with a relatively low CEC, sulfate ions still have a better chance of leaching than cations. Sulfates leaching through the soil can also contribute to micronutrient deficiencies because they often bond with many trace mineral ions. Microorganisms create sulfate by oxidizing reduced forms of sulfur such as sulfur dioxide, various sulfides, and elemental sulfur, and by mineralization/oxidation of organic sulfur sources. As sulfur is oxidized by microbes into sulfate, some enters the soil solution where it is most available to plant roots. Unfortunately, given the way greens are designed to drain, the mass flow of soil solution is usually down and away from the root zone. Additionally, if the green is constructed of calcareous sand, the abundance of calcium ions can bind with sulfate to create gypsum, a relatively insoluble mineral. Turf plants growing on a sand green depending on less than 1% organic matter to deliver adequate S without other direct or indirect sources may eventually suffer a deficiency. Symptoms of a sulfur deficiency include chlorosis, stunted growth, and underdeveloped shoots. The reduction of photosynthesis from a sulfur deficiency may also suppress root development, decreasing drought tolerance.

Even if there is ample sulfur in the soil, if the green, tee, or fairway is compacted, there may not be enough available air in the soil for microbial oxidation of reduced sulfur forms into sulfate. As bacteria and a few other organisms create sulfate ions, they consume oxygen from the soil. If there are inadequate levels of oxygen, fewer sulfate ions can be created but other problems may arise too. The consumption of oxygen where little is available can lead to anaerobic soil conditions and not only produce phytotoxic substances but also stress plants and suppress vital biological functions. Anaerobic conditions may also lead to denitrification of available nitrate dissolved in the soil solution and the reduction of sulfate into hydrogen sulfide, which easily combines with metal ions and forms what is commonly known as black layer.

Many superintendents use elemental sulfur to lower the soil's pH, especially on greens where they are trying to provide optimum conditions for bentgrass or other species of turf. This practice provides more than adequate resources for sulfur-oxidizing bacteria but may be futile for adjusting pH. Greens built and constantly topdressed with calcareous sand may require impractical amounts of elemental sulfur to substantially lower the soil's pH. Calcareous sand can contain more than 20% calcium carbonate, which means that a single inch of sand contains the equivalent of nearly a ton of pure lime per 1,000 ft^2. Multiply that times the depth of sand in the root zone and it's easy to see how difficult pH reduction will be. The amount of elemental sulfur needed to effectively lower the soil pH can cause many more problems than it solves.

A rich soil containing both friable and stable humus naturally buffers extremes in pH. It also provides storage for organic forms of sulfur and provides habitat for the organisms needed to transform it into plant usable sulfate. Biological activity in a rich soil aggregates soil particles, forming quasi-permanent pore space that can protect the soil from anaerobiosis. They do this more for their own survival than for the welfare of plants. Earthworms make passages through which air, water, and plant roots can travel.

Their tunnels connect and widen the interface between the soil and atmosphere, ensuring adequate release of carbon dioxide for plants and ample absorption of oxygen for roots and soil organisms. Regular contributions of well-made, well-aged compost can not only provide organically bound sulfur to the soil but also an army of organisms that efficiently regulate its availability. It is yet another example of how well the natural ecosystem can function if ample resources are available.

Nitrogen, phosphorus, potassium, calcium, magnesium, and sulfur are considered macronutrients because plants use much greater quantities of them than the seven other soilborne elements considered essential. These other elements are aptly called trace elements, trace minerals, or micronutrients because of the small amount plants need to survive. In fact, doses greater than trace amounts can be toxic to plants.

When scientists began testing different nutrients to determine which are essential for plants and which are not, they found that plants could not survive if one of 16 different elements were missing. These elements are carbon, oxygen, hydrogen, nitrogen, phosphorus, potassium, calcium, magnesium, sulfur, iron, manganese, copper, zinc, molybdenum, boron, and chlorine. Logically, the conclusion was drawn that these elements are the only nutrients that are essential to the plant's survival. However, there may be other elements that are essential to plant health. Experiments have shown that many other elements enhance plant growth or vigor but because the plant can survive in their absence, they are not considered essential. In fact, plants can substitute some nonessential elements for essential ones when the need arises. Sodium, for example, can function like potassium in plants and in the absence of K, plants are able to survive using sodium instead. A similar substitution has been observed between vanadium and molybdenum.

Chemical analysis of different plant tissue specimens revealed 26 other elements (in addition to the 16 considered essential). These include antimony, arsenic, barium, beryllium, bismuth, bromine, cadmium, cerium, cesium, chromium, fluorine, gallium, gold, iodine, lead, mercury, nickel, selenium, silver, strontium, thallium, tin, titanium, tungsten, uranium, and vanadium. Whether any of these extra elements contribute to the plant's welfare has not been determined but it is not unreasonable to assume that if the plant absorbs them, then they play some role in one or more of the plant's functions. Some of the elements on this list may seem downright lethal but in trace amounts they can be functional and beneficial. Deficiencies of one or more of these obscure minerals may result in symptoms that betray detection of the problem. The underlying cause of many turfgrass problems may simply be that the plant's defense system is functioning below its potential. This theory is unsubstantiated by research and might be difficult to prove but beg the questions:

- If even traces of almost half the naturally occurring elements on earth are found in plant tissue, is it reasonable to assume that the majority do not contribute to the functionality of the plant just because we are unable to determine how?
- Is it possible that other beneficial elements are also in plant tissue but in such trace amounts it is below the detection limits of our lab equipment?
- And is it likely that even more naturally occurring elements would be found in plant tissue if they were available to the plant's roots?

Silicon, for example, does not appear on the list above but there is plenty of evidence that it benefits some plants. Rice is a crop that responds very favorable where silicon is available. Not only does the plant stand more erect but also appears to have a measurable increase in disease resistance and tolerance to iron and aluminum toxicity. Might silicon have a similar effect on turf? At one time in earth's history, these obscure elements were much more available in the soil than they are today mainly because countless generations of agricultural plants have extracted them from the soil and the elements have not been replaced. The fact that crops absorbed these elements from the soil suggests that they were either required or desired. Some managers believe that soils should periodically be enriched with materials containing a broad diversity of natural elements. Seaweed, kelp, (Acadian, North American Kelp) and many rock dusts from volcanic deposits (Peak Minerals, Summa Minerals) contain a broad diversity of natural elements in trace amounts. These products can be mixed with topdressing materials, compost, or applied alone.

Dried seaweed may be too coarse for direct application to greens. Mowing with the baskets off will chop it into finer particles but, if there is any sand in the seaweed, it can dull or damage the mower. A more practical method is adding it to the compost pile where biological activity will reduce its particle size. Clean, dried, and pulverized seaweed can be added to the topdress mixture. Golf courses in littoral regions may have free access to this resource. Many shorelines are littered with seaweed detritus, especially after a storm, and some municipalities go to great expense to remove it from the beaches. Generally, it is not difficult to obtain the permits necessary for harvesting seaweed that has washed ashore. Some municipalities may be willing to deliver seaweed from beaches directly to the course rather than pay tipping fees at the local landfill. Part of coauthor Luff's maintenance crew used to head for the beach with a dump truck for two or three loads of seaweed once or twice a year. Upon their return, the seaweed would be incorporated into the most active compost piles with a bucket loader. Now, the town delivers seaweed to save on landfill fees.

Rock powders from volcanic deposits are normally a dust-like consistency and can easily be worked or watered through a bentgrass or other tightly-knit turf canopy but they can be difficult to apply through conventional equipment. Drop spreaders are more practical than broadcast spreaders that often create clouds of dust. This swarm of suspended particles can drift away from the area where the application is needed and intended. A drop spreader, however, is a laborious way to cover large areas of turf and any moisture picked up from the turf can quickly gum up the works. Most of these materials are easier to manage if used to enrich compost or other topdress material. Some manufacturers produce a soluble extract from seaweed that is normally applied as a liquid. Seaweed extract is compatible with most other sprays and can counter the stressful effects of some pesticides. It can also be used through the course's irrigation system. The soluble extract actually contains just as diverse an inventory of elements as the dried algal remains and, because the contents of the extract are dissolved, utilization is often significantly greater.

Applications of organic fertilizers, compost, or both may indirectly increase the availability of some obscure elements by increasing biological activity in the soil. Most soils contain more than adequate amounts of trace elements in tightly-bound, mineral formations and the acids, enzymes, and chelates created by soil organisms can dissolve the bonds enough for some elements to become available. Iron, for example, is very abundant in a typical mineral soil but its availability depends on organisms that can free it from its insoluble relationship with the parent material. Some organisms in the root zone can produce a scavenging chemical molecule (called *hydroxamate siderophore*) that finds and secures iron ions for plants. Some strains of mycorrhizae will produce this molecule and retrieve iron for the roots to which they are attached. It is unlikely that this will occur to a great degree on a sand green as the diversity of minerals in sand is relatively low and the degree to which they are insoluble and unavailable is relatively high.

Organic matter also plays a major role in the storage and availability of micronutrients. Its ability to establish complexes with metal ions is well known. The elements in these complexes are more easily accessed by the actions of organisms and even plant roots to a certain extent. Contributions of organic residues from compost, clippings, or organic fertilizers can increase the production of humic and fulvic acids. These acids can not only dissolve micronutrients from their parent particles but can also maintain many of them in soil solution, where they are most available to plant roots. Humic and fulvic molecules often complex with metal ions, forming chelates that prevent the reestablishment of mineral bonds but allow plant and microbial access to the nutrients.

If the health of soil organisms is directly related to the health of turf, then it is logical that nutrients important to them are also important to turf plants. Turf plants may not absorb some of these elements and they are therefore regarded as unimportant. But, if these elements are essential for the functions of beneficial soil organisms, then they must also be regarded as important for turf. Soil scientists at the United States Department of Agriculture recognize the fundamental importance of interactions between soil organisms for plant growth and health. Deficiencies of essential nutrients for any of these organisms could easily affect their ecological balance in the soil and the interaction between them. The list of nutrients considered essential for soil organisms is quite similar to the list for animals; however, the methods of measuring essentiality for plants and animals are different. As we previously mentioned, nutrients considered essential for plants are those that enable plants to survive whereas nutrients essential for animals may only serve to enhance growth or benefit functionality. The animal may be able to survive without them. There is only one element, boron, that is considered essential for plants but not for animals but there are several that are considered essential for animals but not necessarily for plants. These include chromium, cobalt, fluorine, iodine, nickel, selenium, sodium, tin, and vanadium. It stands to reason that if the growth and function of soil organisms are enhanced by any of these elements, then plants would also benefit. Cobalt, for example, is essential for symbiotic nitrogen-fixing bacteria. Without cobalt, these essential organisms would not be able to survive and colonize the roots of leguminous plants. Since cobalt is essential for the conversion of atmospheric N into protein, it is reasonable to assume that free-living nitrogen-fixing bacteria need it too.

We are not suggesting that concentrated preparations of any trace element be arbitrarily applied to the soil. Unless a micronutrient deficiency has been determined by a soil test or leaf tissue analysis, the threshold between helpful and harmful is too narrow for these types of amendments. Seaweed, volcanic rock dusts, or both can, however, add a rich diversity of slowly released nutrients to the soil without increasing any one to a phytotoxic level. Decades of research on the use of both volcanic rock dusts and dried seaweed reveal not only the potential for benefits but also the improbability of harm.

Regardless of whether rock powders or seaweed is applied, the wealth of nutrients they contain may remain unavailable to plants if there isn't enough biological activity in the soil to release it. Soil organisms release trace elements through the production of organic acids, enzymes, and chelating agents used to extract nutrients from decaying plant or animal residues and from inorganic mineral sources. Much of what they extract is assimilated and then released for plants when the organisms are preyed upon or when their dead bodies are consumed by saprophytes. There are times when soil organisms will make trace elements less available by immobilization or oxidation. If deficiencies occur from biological immobilization, it is unlikely that the pool of trace elements was adequate before bio-consumption. An affluence of biological activity may usurp the supply of trace elements in the soil but, normally, it would be ephemeral. The army of predators that commonly follow a bloom of decay organisms would invariably release ample amounts of consumed nutrients. Plant roots exude a wide variety of organic compounds that attract and feed a relatively diverse population of soil organisms. But plants have the ability to attract a more specific group of organisms that better serve the plants' needs. Shortages of a particular nutrient may induce the plant to release pabulum that attracts organisms better suited to secure what the plant needs. Oxidation transforms some nutrients into plant-available forms and others into unavailable forms. Biological oxidation is balanced by biological reduction, which contributes to the system of storage, release, and availability of many nutrients.

Growing turf in a relatively inert medium such as sand is a little like caring for a newborn child. It can't tell you what it wants; it can only cry, and its needs are constant. The superintendent has to spoon-feed everything that the child cannot—through a functioning soil ecosystem—provide itself. And those needs are erratic and inconsistent. This morning it was nitrogen, this afternoon, water. Yesterday it was iron and the day before that it had a type of diaper rash called dollar spot. This is not to say that fertilizer is unnecessary on an ecologically maintained golf course but the system of nutrient exchange and cycling, biological oxidation and reduction, and natural storage and release doesn't function well in an environment where too many important components are missing. Unfortunately, adding only soluble nutrients to the baby's formula rarely helps the situation. We've taken the bottle (or the breast) from the child and replaced it with an intravenous drip. Eventually, those unused digestive organs become dysfunctional and the baby's health begins to decline. There are a countless number of components in a whole ecosystem that function spontaneously. To manually provide all the benefits of those chemical, biological, and physical interactions, using only symptoms as a guide, is a monumental task.

Chapter 3:

Compost

A superintendent who is unable or unwilling to use compost on his course may not achieve freedom from chemical dependency anytime soon. Compost is an essential tool on any ecological golf course but, like any other tool, it must be well made to produce quality results. There is compelling evidence that topdressing with mixtures of sand and compost or just compost alone notably improves soil conditions and turf performance. If the compost is made well and aged correctly, it can also suppress disease organisms, reduce thatch, inhibit compaction, diminish palatability of plant tissue to herbivorous arthropods, reduce the soil's thermal conductivity, increase turf's stress tolerance, and decrease the course's water consumption, all to a very significant degree. Because of its potential for inhibiting pathogens, compost is currently being examined as an alternative to methyl bromide (a soil fumigant facing regulatory extinction). Compost not only contains many groups of beneficial organisms but provides resources for indigenous groups of organisms as well. The populations of these beneficial soil-dwelling creatures swell after compost has been applied and as their numbers increase; so do the beneficial tasks that they perform. They can regulate fertility (see Chapter 2, Fertility), for example, and reduce the superfluity or deficiencies of nutrients often experienced using conventional fertilizers. Their regulation of nitrogen reduces the level of free amino acids and nitrates in plant tissue which can act as attractants to insects and pathogens. Compost also provides fertility, which eventually may supplant the need for other fertilizers. For each one of the hundreds—perhaps thousands—of benefits from compost that scientists now recognize, there are likely scores of advantages yet to be discovered.

Making compost with waste materials generated on or near the course may be practical for some superintendents. For others, the task may be unfeasible. There are many reasons why on-site composting may be unrealistic but not having the necessary skills to make compost should not be one of them. Those can be acquired with practice and patience. The time, equipment, and materials needed to make quality compost may not be readily available or within the club's budget. Lack of a suitable location or local ordinances may also pose restrictions. Purchasing compost from private or municipal facilities may be the only alternative but the superintendent must be able to judge quality compost or, more accurately, judge material that has the potential to become quality compost.

Using compost on the golf course is not new. The Turfgrass Information Database at Michigan State University has literally hundreds of articles written in the early part of the twentieth century about making, curing, storing, and applying compost to greens, tees, and fairways. Superintendents back then understood the need for a healthy, func-

The topdressing machine is applying a mixture of fifty percent sand and fifty percent well-aged compost. The mixture holds more moisture than sand alone and requires a little more sweeping to work it through the canopy.

tioning ecosystem without knowing nearly as many of the benefits as are known today. Most superintendents, for example, used compost to dormant-feed turf without knowing that it also offered protection against winter diseases. The use of compost was ubiquitous but relatively little was known about the subtle but important differences in processing and aging techniques. Even though less information about making and using compost was available back then, most superintendents had a fair amount of experience with it. During the chemical revolution that began after World War II, superintendents began to lose touch with the composting skills they had acquired in the preceding decades. The ease with which modern pesticides could control symptoms lured many well-meaning managers away from the basic tenets of soil and plant health. Unfortunately, this new trend nullified the need to pass-on the wisdom of composting to subsequent generations of superintendents. Over the years, while only a few people were paying attention, the art of composting has been refined and a lot more discoveries have been made about its potential benefits. Much of this information was uncovered by accident from research aimed at solving solid waste disposal problems. Now, superintendents are beginning to rediscover the benefits of using compost and, at the same time, learning that well-made compost has much more to offer than their predecessors ever imagined.

Compost undergoes a complex transformation into many different products as it is consumed and recycled by soil organisms; the end product is humus—the most stable component of soil organic matter. However, humus does not support the degree of biological activity that compost does. Organic matter, as a whole, is an essential component of a natural soil system and it would be hard to find any expert who disagrees. The term 'organic matter' describes many different materials from thatch to humus. It also includes senescent roots, shoots, and the bodies of soil organisms. Within each category of residue, there are an infinite number of decomposition stages that all have distinct characteristics. About the only thing these different organic materials have in common is that they are all dead. Every organic residue at all stages of decomposition has something to offer the soil ecosystem. Mature compost is a stage of decomposition that is relatively stable but, unlike humus, still offers a plentitude of resources for soil organisms. Soil organisms improve soil structure by binding organic matter together with clay and other soil particles creating porosity for air, water, and roots to travel through more easily. Organisms aggregate clay by interrupting the bonds between particles. These bonds can be extremely tight and make the soil more prone to compaction and problems associated with air and water movement.

Soil organisms thrive in organic matter. It has everything they need to exist, including water, food, and air. Consequently, many different varieties of microorganisms are constantly competing for those available resources. That competition creates a biological equilibrium that, in the long and short term, can prevent or mitigate many insect, disease, and even nutrient imbalance problems that plague the superintendent. Carbon in organic matter is the source of energy for most soil organisms. The greater the available source of energy, the larger and more diverse their populations can be. Compost, however, has many more resources to offer to soil organisms than humus. As the humification process matures, there are less available nutrients for soil organisms, so fresh material needs to be added on a regular basis to maintain a high level of biological activity.

Most forms of organic matter can hold many times their own weight in water. This property increases the soil's capacity to retain more dissolved nutrients and decreases the likelihood of nutrients leaching away from the root zone. Humus itself is not porous, but its ability to hold as much water as it does can create porosity. Humus swells in order to accommodate as much as four to six times its own weight in water. This swelling causes a sort of heaving in the soil. When water is released during drier periods, organic matter contracts and leaves behind air spaces. Earthworms and other larger organisms that burrow through the soil looking for organic residues create even more porosity. This activity lightens the soil but does not necessarily create a soft playing surface. Humus—the most mature and often the most abundant form of soil organic matter—is also a colloidal substance that can magnetically hold mineral nutrients such as potassium, magnesium, ammonium, and calcium ions. Nutrients such as nitrogen, phosphorus, and sulfur exist as proteins and other compounds in organic matter, and can be mineralized into available nutrients by soil organisms at a slow, sustained, and efficient rate.

Organic matter is lighter than sand, silt, or clay, which benefits root growth. As roots travel through the soil, energy, manufactured by photosynthesis, is used. The more difficult the journey, the more energy the root uses to travel the same distance and root growth is often stunted. The farther roots travel and branch out through the soil, the more access they gain to nutrients needed for synthesizing more energy.

Mineral nutrients are made available in the soil from the activities of organisms on parent material (rock particles). There are approximately 52,000 pounds of potash, 40,000 pounds of magnesium, 72,000 pounds of calcium, 3,000 pounds of phosphate, 3,000 pounds of sulfur, and 100,000 pounds of iron in an average acre of topsoil (see Figure 1.3). Albeit abundant, the vast majority of these nutrients are bound in mineral structures that are insoluble and unavailable to plants. There are many organisms in nature that can dissolve some of these minerals into available nutrients for plants. Soil organisms release acids and chelates that free nutrients from the surface of soil particles. They produce corrosive exudates specifically to liberate mineral nutrients for their own metabolisms. Once assimilated by an organism, the mineral is bound in an organic complex that is more readily available to other soil life after the organism dies or is consumed by a predator. The mineral can be recycled endlessly through all different types of soil organisms, including plants. Organic matter, especially from mature compost, directly influences the existence and proliferation of organisms that can free minerals from the soil's vast reserve.

Most scientists will admit that for every known benefit from compost, there are several that are unknown or unexplainable. Recently, a phenomenon referred to as compost-induced systemic-acquired resistance was discovered—a fancy name for the unexplainable disease resistance systemically spread throughout a plant growing in a compost-based media. Researchers, using containerized plants, infected half of the plants' roots—separated from the other half by an impenetrable divider—with disease pathogens. The infected half of the roots were growing in a peat mix. The uninfected half of the roots were growing in peat, soil, or a compost-based mix. The containers where the latter half of the roots were growing in either peat or soil showed symptoms of the disease on the roots in both sides of the container, and on the aboveground portion of the plant. The containers that had half the roots growing in a compost mix showed disease suppression in all parts of the plant, including the side of the container where the pathogens were introduced. Scientists could not explain the phenomenon but they could clearly conclude that it occurred. (W. Zhang et al.).

Another important benefit of applying compost to greens, tees, and fairways is the generation of carbon dioxide. Carbon dioxide cycles from the atmosphere into plants and other photosynthesizing (autotrophic) organisms. Proteins, carbohydrates (sugars and starches), fats, waxes, lignin, cellulose, hemicellulose, and other carbon compounds produced by autotrophic organisms are food for almost all other living things on earth, including soil organisms. Each time energy is extracted from these carbon-based foods, carbon dioxide is produced and released back into the atmosphere (see Figure 1.8). Plants somehow know the value of biological activity on and around their root zone and release, through the roots, up to 50% of the photosynthesized foods that they produce to feed soil organisms. In a perfect system, all the carbon dioxide that is produced by organisms existing on food released from roots would be reabsorbed by plants. Unfortunately, that doesn't occur and supplemental carbon dioxide is needed. The most likely source is from the decomposition of organic residues.

All of the nitrogen, phosphorus, potassium, calcium, magnesium, sulfur, and trace elements that plants use comprise only 5% of their diet. The other 95% is from carbon, oxygen, and hydrogen. Plants derive carbon, hydrogen, and oxygen from carbon dioxide in the atmosphere and water. Organic residues decaying in the soil produce carbon dioxide that seeps up to the soil surface where plants can absorb it. And even though there is growing concern about the buildup of carbon dioxide in the atmosphere—and its relationship to global warming—there are often inadequate levels of it for plants. In a natural system, the constant decay of plant and animal residues on or near the soil's surface produces a constant source of carbon dioxide for plants. The sand-based systems on modern golf course greens have few organic residues to offer decay organisms and where the clippings are collected, there are even less. The inner atmosphere of the earth occupies about 100 miles of space over the planet whereas grass plants often occupy less than an inch (on greens, less than ¼ inch). It is difficult for the carbon dioxide that is produced elsewhere to settle on the golf course for any length of time. Although carbon dioxide is slightly heavier than air, forces such as convection currents, trade winds, prevailing winds, jet stream, pressure gradients, sea breezes, and land breezes tend to move carbon dioxide into portions of the atmosphere that are inaccessible to plant leaves. There are moments—usually in the early morning or evening—when air currents are relatively calm. Coincidentally, these are the same moments when the plants' stomates (pores on the underside of leaves) are open, allowing for the exchange of gases. It's unlikely that these phenomena coincide accidentally. If carbon dioxide is being generated in the soil, plants rooted there can absorb it as it seeps from the soil's surface. Compost offers many different forms of food for soil organisms ranging from bacteria to earthworms, all of which derive energy and respire carbon dioxide that can be absorbed by plants.

The list of benefits from adding compost can go on for several hundred pages, and the expense of providing those benefits without regularly contributing this organic matter is staggering. It is important not only to add compost on a regular basis but also to limit activities that deplete organic matter in the soil (see Chapter 6, Cultural Practices). Routine cultural practices performed by the superintendent and his crew may inadvertently deplete soil organic matter content. When organic matter levels begin to drop, so do the corresponding benefits, including water and nutrient holding capacity and the production of carbon dioxide from respiring populations of soil organisms. This loss has a profound effect on the soil, creating an economic recession that impacts the entire soil ecosystem. The only things that are often increased by lower levels of organic matter are bulk soil density and problems.

Making high-quality compost is more an art than it is a science. Learning the most practical method will take some effort and time but, like anything else, practice provides proficiency. Keep in mind that the decay of organic matter is a natural phenomenon that will occur with or without our help. All we are trying to do is speed up the process, create heat to kill weed seeds and pathogens, and produce a product that will not only improve the health and appearance of turf but protect it as well. It is very important to understand, however, that all composts are *not* equal. The difference between high-quality and low-quality compost is comparable to the dissimilarities of a vintage Rolls Royce and a pile of scrap metal.

There are five criteria for accelerating the decay process to successfully produce compost in a reasonable amount of time. These conditions are adequate air and water, correct carbon to nitrogen ratio (C:N), initial ambient temperatures above freezing, and a pH range between 5.5 and 9.0.

The organisms that are most adept at breaking down organic residues are aerobic; they need oxygen to live. If a pile of material is wet and compressed, there will not be enough air for the right type of organisms; instead, another type, the anaerobe will process the pile. This condition can cause odor problems, will not create the heat necessary to kill weed seeds or pathogens, can produce phytotoxic compounds, and does not produce the same end product—one that is capable of providing disease suppression and feeding appropriate groups of soil organisms. In fact, poorly made compost can actually increase the severity of disease. Turning and aerating the material several times increases the amount of oxygen in the pile and moderates the temperature. Many compost piles are set up on top of perforated pipes to ensure that plenty of air is reaching the bottom of the pile (see Figure 3.1). There is a theory that proper mixing and a piped air infiltration system are all that is needed for proper aeration. The heat generated by the pile creates a chimney effect that draws fresh air in through the pipes. No turning or motor-driven blowers are necessary, but creating a porous mix to begin with is crucial. If the mix gets compressed by precipitation or just by gravity, air movement may be inhibited and so will aerobic decay. If entrainment loss is severe, the pile may become anaerobic. During conventional composting, turning is a crucial component.

FIGURE 3.1. GRASS PILE
Temporary Storage

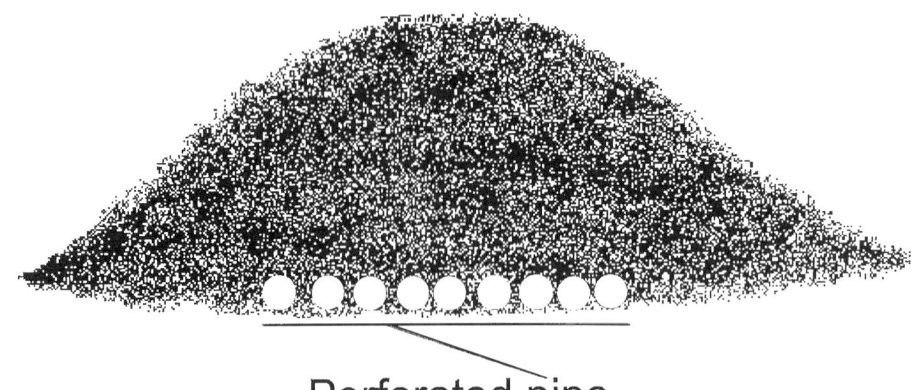

Perforated pipe
(length should reach through pile base)

Knowing when and how often to turn a pile is the key to making compost quickly. Unfortunately, there is no set formula. The number and frequency of turns necessary changes with the climate and the different materials being decomposed. A good method of determining turning frequency is by using a thermometer with a two to four foot long probe to monitor the temperature deep within the pile. When the pile is first mixed the temperature should rise within one to three days to between 120° and 160° F. When the temperature drops 20° (F) below its peak, it is a good time to turn the pile. It is also a good time to turn if the pile gets too hot (>160° F). Each time the pile is aerated, heat will increase again until the process is nearly completed. At the point where the temperature is relatively stable, the compost is ready to cure. Eventually, familiarity with the composting process can eliminate the need for a thermometer. As mentioned earlier, compost happens whether we participate in the process or not. Many people who make compost don't have time to turn the pile regularly and, consequently, the process is retarded. But, eventually, compost is made and it can possess the same characteristics as the compost that was pampered; it just takes longer. Turning frequently at the early stages of the compost process ensures good aerobic decomposition, creates enough heat to kill weed seeds and pathogens, and shortens the amount of time needed for the pile to cure. Coauthor Luff's crew turns their compost piles once every two weeks on average. Idealistically speaking, more frequent turning during the early stages of composting would be preferable but a superintendent's time can often be the most limiting factor.

In some large composting operations, windrow turners are used to mix and aerate the pile. These machines are specifically designed to add air into the windrows and circulate all materials into the core—the hottest part of the pile. For a golf course making just enough to satisfy its own needs, a bucket loader is a quick and efficient way to turn a pile. When turning the pile with a loader, scoop up the compost, raise the scoop to an elevated position and allow the material to cascade slowly from the bucket. Generally, the pile is moved each time it is turned, so a site that is at least twice the size of the pile's footprint (plus maneuvering space) is ideal.

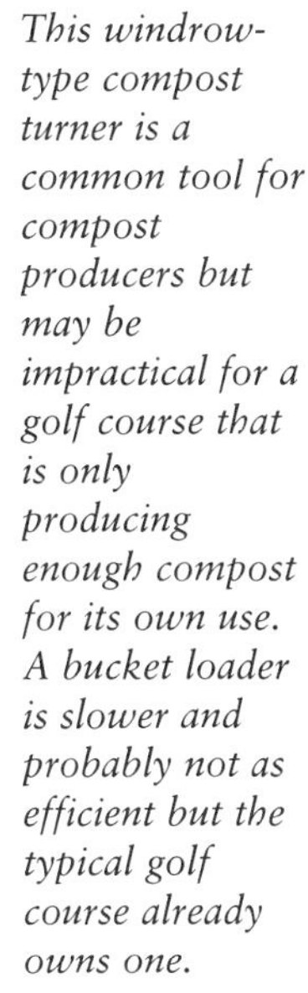

This windrow-type compost turner is a common tool for compost producers but may be impractical for a golf course that is only producing enough compost for its own use. A bucket loader is slower and probably not as efficient but the typical golf course already owns one.

Surface area is another aspect of aeration that needs to be considered. The smaller the particle of organic material, the more surface area is exposed to decay organisms, and the faster it is decomposed. Sawdust decays faster than chips, which decompose considerable faster than whole branches. A proper mixture of coarse and fine materials also helps air infiltration. Material that is too fine can decrease air infiltration through the pile and retard aerobic decay or cause anaerobic conditions. The greater the amount of surface area, the faster the microbial consumption of oxygen, and the more often turning is needed.

Moisture is an important consideration in the compost-making process. All living things need water to live, and the organisms that make compost are no different. A compost pile should contain somewhere between 40 and 60% moisture. A simple test to determine proper moisture content is to squeeze a handful of raw material. The material should feel damp to the touch but no more than a couple of drops of water should be expelled. If the pile is too dry, reproduction of the composting organisms can be inhibited, which can slow down the composting process. If the pile is too wet it can easily become anaerobic. During the composting process, the moisture level in the pile can vary. Heat and turning can drive off moisture whereas excess precipitation can soak a pile and force out too much oxygen. Sometimes a pile can get too hot if there is inadequate moisture because the evaporation of water moderates the pile's temperature. Adding water to a pile is easy enough if it is needed, but extracting it is impractical. To correct excess moisture content dry, bulky materials can be mixed into the pile to absorb the water. Saw chips or sawdust, planer shavings or shredded paper can be used. Paper products tend to mat when wet so thorough mixing is necessary. Newspaper is more likely to mat and decomposes more slowly than other uncoated papers. Clay-coated papers used to produce glossy printing are not recommended. Dry leaves are suitable if it is the season for them, and soil can also be used in a pinch. Many of these dry, bulky materials can alter the carbon:nitrogen ratio of the pile if too much is incorporated. If a wet pile fails to reheat after adding moisture absorbing material, the C:N ratio of the pile may have become too wide and feedstock with a narrow C:N ration may have to be added. Turning the pile can also help to dry it out.

The ratio of carbon to nitrogen (C:N) is important when combining materials for composting. The ideal C:N ratio for a compost pile is between 25 and 35 parts carbon to one part nitrogen (average 30:1). If the C:N ratio of the pile is too wide (i.e., high carbon, low nitrogen), composting will be slow because the amount of available nitrogen is too low for proper protein syntheses, and the reproduction of decay organisms is inhibited. Adding organisms or enzymes in the form of compost starters or inoculants will not correct this problem if the ratio is very wide. If the C:N ratio is too narrow (i.e., too much nitrogen, too little carbon), reproduction can be overstimulated, causing rapid oxygen depletion and eventually, anaerobic conditions. If there is insufficient carbon for microbial processing, excess nitrogen can be lost to leaching or volatilization. If the pile begins to generate unpleasant odors, chances are good that nitrogen is being lost through volatilization. Estimating how much of what material to add to the pile can be accomplished by using some of the values in Table 3.1.

These values (Table 3.1) are in ranges because the C:N ratio of organic residues varies depending on age, variety, where the organic residues came from, or what nutrients they absorbed when they were living. C:N values are calculated on a dry matter (weight) basis, but compost feedstock is usually measured volumetrically. Additionally, the bulk density of these different materials varies greatly. Residues such as clippings, green leaves, garden wastes, food wastes, manure, and even soil generally have a more narrow C:N ratio and are usually heavier—often because they contain more water. Residues such as dried leaves, wood chips, and straw usually have a wider C:N ratio and are lighter. Calculating exactly how much of each component to use must be done by dry weight. If, for example, equal (dry) weights of two different materials are mixed together, one with a C:N ratio of 100:1 and the other with a ratio of 10:1, the resulting ration is 110:2 or 55:1. To calculate this accurately, the moisture content must be determined first. This step seems both complex and impractical. Fortunately, it is also unnecessary. Table 3.1 gives a rough idea of what to use to either raise or lower the C:N ratio of a pile. The ideal level for fast decomposition is around 30:1, so a combination of materials that have values above and below that level are a good starting place. A good

Table 3.1. Carbon to Nitrogen Ratios

Material	C:N Ratio	Material	C:N Ratio
Alfalfa meal	15	Linseed meal	8
Animal tankage	7	Newspaper	400–850
Apple pomace	48	Oat straw	50–100
Aquatic plants	15–35	Paper	125–180
Blood meal	3	Paper fiber sludge	250
Cardboard (corrugated)	560	Paper mill sludge	55
Castor pomace	8	Paper pulp	90
Cocoa shells	22	Paunch manure	20–30
Coffee grounds	20	Pig manure	10–20
Compost	15–20	Potato tops	28
Corn silage	35–45	Potatoes (culled)	18
Corn wastes	60–120	Poultry manure	5–15
Cottonseed meal	7	Rice hulls	110–130
Cow manure	10–30	Sawdust	200–750
Crab/lobster wastes	4–4.5	Sawmill waste	170
Cranberry wastes	30–60	Seed meals	7
Fish wastes	2.5–5.5	Sewage sludge	5–16
Food wastes	14–16	Sheep manure	13–20
Fruit wastes	20–50	Shrimp wastes	3.5
Garden wastes	5–55	Slaughterhouse wastes	2–4
Grass clippings	9–25	Softwood bark	100–1000
Grass hay	32	Softwood chips, shavings, etc.	200–1300
Hardwood bark	100–400	Soil	12
Hardwood chips, shavings, etc.	450–800	Soybean meal	4–6
Hoof and horn meal	3	Tree trimmings	16
Horse manure	22–50	Turkey litter	16
Leaves	40–80	Vegetable wastes	11–19
Legume hay	15–19	Wheat straw	100–150

beginning mix is approximately three parts of the lighter, more carbonaceous material to one part of the heavier, more nitrogenous residues (by volume). Depending on actual ingredients, particle sizes and moisture content, adjustments may be necessary. Like anything else, experience makes it easier.

If the majority of available wastes have a wide C:N ratio (and a slow composting rate), organic or inorganic nitrogen fertilizer can be used to create a better balance. Fertilizer, however, should be used very judiciously and mixed in thoroughly. Remember, if a little is good, more is not necessarily better. As soon as odors become evident, chances are that too much nitrogen has been applied. Carbonaceous additions and turning (oxygenation) are necessary to remedy the situation. It is also important to understand that decay organisms need more than just carbon, nitrogen, oxygen, and hydrogen. They, like plants and animals, need many other nutrients to thrive and proliferate. Deficiencies of any essential nutrients will limit population growth. Most residues contain a broad diversity of nutrients but some materials, like wood chips or sawdust, contain mostly carbon.

Reducing particle size of the waste by grinding or shredding can help speed up the composting process in a pile that has a C:N ratio wider than 30:1. Smaller particles afford more overall surface area for composting organisms to decay. There is a limit, however, to how much this will help a pile heat up if the C:N ratio gets too wide. There is also a limit to how small particle sizes should be. Piles of very fine particles have less porosity and can contain too little oxygen for proper composting. If strong, unpleasant odors are evident each time the pile is turned, then the pile is probably using oxygen faster than it can be replaced.

Another important consideration regarding compost ingredients is the groups of soil organisms that the superintendent wants to proliferate on a golf course. Most turf performs best in a soil that is slightly dominated by bacteria or has roughly an equivalent ratio of active bacteria and fungi. Composts made from mostly woody material, such as dry leaves/needles, bark, wood chips, or straw are high in cellulose and lignin which support fungal organisms more than bacterial. Even after these materials are completely decayed, the populations of different fungi tend to outweigh the populations of bacteria. Compost that is ideal for a golf course can be made with some woody material but a substantial amount of material high in carbohydrates and proteins such as clippings, green leaves, food waste, or manure needs to be incorporated. Molasses, syrups, table sugar, plant extracts, or some other type of natural sugar can be added to increase populations of bacteria (for a list of other suitable materials, see SFI 2000a in Sources and Resources). These materials ensure that good populations of beneficial bacteria are cultivated and the compost, after proper heating, curing, and aging, will contain the highest level of disease suppressive characteristics. If the superintendent is buying compost from an outside source, he should inquire from what feedstock the compost was made and should avoid buying compost made exclusively from woody yard wastes. This is not to say that yard waste compost is detrimental or toxic to turf, but it may not be the most ideal in terms of peak turf performance and disease suppression. If yard waste compost is the only kind that is available, mixing and curing it with one or more of the sugars mentioned above can significantly increase its potential for disease suppression. The material should be bioassayed, however, before it is used for disease prevention or control. Many producers of yard-waste compost add ingredients such as grass clippings, manure, or garden wastes that significantly change the biological balance of the finished product. Bio-analysis can indicate its potential for disease suppression.

Curing compost may require turning the pile occasionally to see if it will heat up again but often it just has to sit. The time it takes to cure a pile is debatable but research at Cornell University suggests that if the superintendent wants disease suppressive characteristics in their compost, the pile needs to cure for at least two years. This means the superintendent has to plan for the future. That planning includes calculating how much compost will be needed. The ratio of sand to compost in a topdressing mixture can be as rich as 1:1 (sand to compost) but less than 7:3 may be inadequate. Straight compost has been used on greens over the winter months to protect them from snow mold; however, much of the compost needs to be removed in the spring. Cornell University experimented with a half inch layer of compost (200 lb per 1,000 ft^2) on greens over the winter to successfully combat gray snow mold but had to remove it in the spring before the grass resumed growth. To cover an acre with a quarter-inch thick layer requires ~34 yards of a compost/sand mix. Calculating mixture ratio, coverage area, and application frequency will give the superintendent a rough idea of how much compost he will need to stockpile.

To calculate how much material is needed to cover a given area, multiply the thickness of the intended topdress layer by 3.086 for yd^3 per 1,000 ft^2, or multiply by 134.44 for yd^3 per acre. For example, to topdress a 3⁄16 (0.1875) inch layer of material,

A curing compost pile is seldom turned and, consequently, can host many plants from wind-borne seed. Its outward appearance belies the value within.

$$3.086 \times 0.1875 = 0.58 \text{ yd}^3 \text{ per } 1{,}000 \text{ ft}^2\text{, or}$$

$$134.44 \times 0.1875 = 25.2 \text{ yd}^3 \text{ per acre.}$$

If the topdress mixture contains 50% compost, then 12.6 yd^3 of compost is needed per acre of turf per topdress application. (To convert fractions to decimals, divide the numerator by the denominator; e.g., ¼ = 1 ÷ 4 = 0.25.)

Compost that is available from private or municipal compost making facilities needs to be examined and tested before it is purchased. Two quick ways to test for quality is to feel it and smell it. Dig deep into the pile and pull out a sample. If it is hot, has objec-

Mature, well-aged compost has a rich look, feel, and smell. It is well worth the time and effort it takes to produce a quality compost like this.

tionable odors, or both, it is probably still immature and needs more turning before it can begin curing. The superintendent needs to determine whether he has the time, space, and equipment to finish the process. It may be necessary to make accommodations for such an operation if a more finished product is not available elsewhere. Most commercial compost makers do not produce finished compost to the degree that it is being described here because the market absorbs most of it before it can age long enough. It is unlikely that the superintendent will find commercially available compost that is even one year old, let alone two to five years old or compost that is made and cured carefully enough to provide disease suppression. Immature compost may be advantageous because it enables the superintendent to control the final curing process. Some commercial producers begin the curing process too soon and the compost pile can become anaerobic. Once this happens, the chemical and biological composition of the compost changes. Curing will only reverse these changes if there is enough porosity for air to flow passively through the pile. If severe anaerobic conditions persist, phytotoxic compounds may be produced that can not only diminish turf performance but could also stress plants—increasing their susceptibility to disease, decrease their competitiveness with weeds, and reduce their tolerance to herbivorous insects. If compost has become anaerobic, aerobic organisms need to be reestablished by adding air to the pile. Frequent turning and possibly the addition of some carbonaceous residues may be necessary to reverse the anaerobic process. A pile that is ready for curing should not reheat by an appreciable amount after it has been turned. If immature compost is purchased, the superintendent can continue turning the pile periodically until reheating ceases to occur. At this point, the superintendent can feel relatively confident that anaerobic conditions will not prevail and that the pile will cure adequately for the next two to five years before it will be used.

Each year the superintendent needs to plan for his compost needs two to five years in the future. It is best if more compost is acquired than the anticipated need because of shrinkage and because it's a good idea to have leftover compost. The volume of compost purchased can shrink as much as 30% depending on the degree of its immaturity at delivery time and leftover compost, like a fine wine, gets better with age. As leftover compost approaches five years of age, it can make excellent compost tea that can be applied to turf areas afflicted with disease symptoms (see Chapter 5, Pests). The art of making good compost tea is another skill that the superintendent may have to master if he plans to part with fungicides anytime soon.

Compost tea has been used for centuries to enhance crop performance but it is only recently that strong disease suppressive characteristics have been discovered. Apparently, most of the organisms that are delivered to plant leaf surfaces from compost tea parasitize, antagonize, or just plain steal food from disease organisms to the point where they can no longer compete. Tea that seeps into the soil protects the roots in much the same way. Making effective tea, however, requires as much attention and care as making good quality compost. In fact, if the compost, from which the tea is to be derived, is not made from ingredients rich in carbohydrates with the correct C:N ratio, turned or aerated correctly, moisturized adequately, cured, and aged well, there is little hope that the tea will display strong disease suppressive characteristics. Unlike the tea that is served with crumpets, compost tea is not steeped in hot water until the color of the water darkens. Cool, nonchlorinated, and oxygenated water is used. Commercial brewers are available that ensure the survival of aerobic organisms in the tea (Growing Solutions, Northwest Irrigation). When compost is placed in a brewer, it cannot be packed too tightly or the

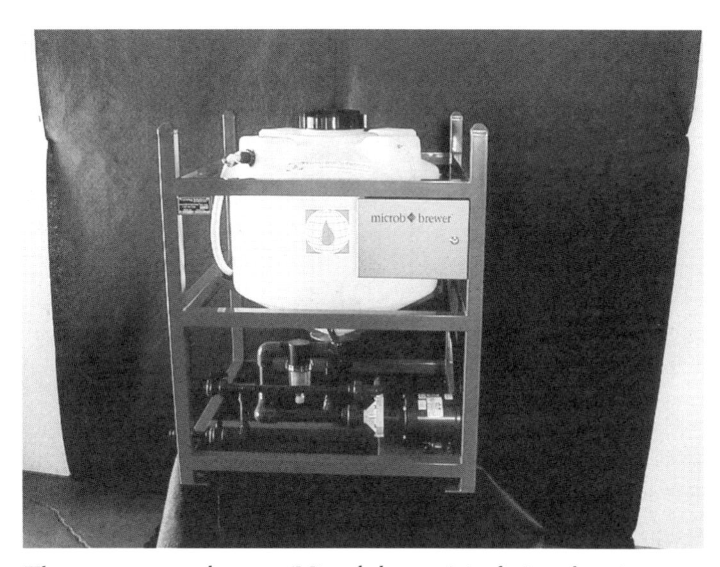

The compost tea brewer (Microb-brewer) is designed to incorporate oxygen into the water so that aerobic organisms can thrive. Compost tea made without oxygen can produce anaerobes which may be harmful to turf.

water will not flow through it adequately and inadequate numbers of organisms will be extracted. If the compost is too wet at brewing time, there will likely be a microbial imbalance in the tea. The squeeze test—where you squeeze a handful of compost as hard as you can—should yield no more that a drop or two of water. The sniff test is important too. If you smell any foul or acidic odors, chances are good that the tea will be ineffective. These may not sound like highly analytical testing procedures but it's truly amazing what experienced senses can perceive. There should be a slight temperature increase (five to ten degrees) in the water during the brewing cycle. Reproducing organisms generate heat but this is sometimes difficult to notice if ambient temperatures are too cold or too warm. When applying compost tea, back off on the pressure a little. Most organisms can withstand high pressure but when they hit the leaf surface or ground at excessive velocities, it's a little like dropping a pumpkin from a second story window. Presently, the thought of making good compost and effective compost tea may seem daunting but keep in mind, so was tying shoes at one point. There's no doubt that anyone who's worn laced shoes for any length of time has gotten the hang of it by now. So it will be with the production of compost and compost tea.

Good quality compost tea is not something that is likely to be commercially available. It has to be made with enough agitation in the container to ensure adequate oxygen levels in the water or the valuable aerobic organisms will be destroyed and the possibly destructive anaerobic organisms will be cultivated. Additionally, compost tea must be used as soon as possible after production or the aerobic organisms will die and anaerobic organisms will grow. Twenty pounds of compost can yield 50 gallons of tea and will cover two to ten acres. The ideal application of the tea is a fine mist with small droplets that adhere to the plant leaf surfaces but, if that's impractical, it can also be injected into an irrigation system and used as a drench. This method has also shown excellent disease suppression. Research of compost tea on turf for disease suppression is in its beginning stages and, since it's unlikely that it will ever become commercially available, little funding is available for testing. However, there is testimony from many superintendents who have tried it and there are many research papers with positive evidence that compost tea can suppress disease on many types of plants.

If the superintendent has only enough time, space, or resources for a small compost pile (< 1 ton of finished compost), then the best use of it would be as a tea. Twenty pounds of compost can make enough tea to treat two to ten acres. Remember, the compost and the tea have to be made right to be disease suppressive.

Once the value of a mature, well made compost has been discovered, there will be a compelling tendency for the superintendent to use more of it. This can present a supply problem especially if the compost is being made or cured on-site. The availability of space or a suitable site may be a limiting factor. Local zoning regulations may also present a problem. If compost is made on-site strictly from organic wastes generated on the course, the quantity of available wastes may be inadequate.

Available space or suitable sites may seem like difficult obstacles to overcome. Neighbors may not appreciate conspicuously placed piles of organic debris and golfers would probably snarl at the idea of sacrificing any in-bound playing area for the production of compost. Add to these conflicts the bureaucratic hoops that one may need to jump through in order to obtain necessary permits and the idea of using compost becomes more daunting. However difficult the situation may seem, the clear and immediately apparent benefits from using compost can serve to increase a superintendent's resolve to overcome spatial and regulatory hurdles. In the face of growing intolerance to horticultural chemicals, most zoning or permitting agencies would favor plans to reduce or eliminate their use, and there is plenty of scientific evidence that mature compost will help accomplish this goal. If compost is produced strictly from on-site residues, it is unlikely that permits will be necessary.

Groves of trees can make ideal sites for compost piles. They are usually out-of-bounds and the shelter trees provide can protect the pile from the sun and wind. If there is enough low-growing brush around the outer edge of the grove, the compost pile will also be out of sight. The grove needs to have a center area with maneuvering space for a bucket loader so the piles can be turned. Some embankments are often ideal for organic waste disposal but impractical for turning compost. On the other hand, if a loader can access the bottom level of the embankment, it may be an ideal site.

Generating enough organic waste to produce ample amounts of compost may be a concern even if space, location, or permits are not. Grass clippings alone will not produce enough compost to treat the greens, let alone the entire course. Coauthor Richard Luff conserves his compost by using it only on the greens but we have no doubt that the entire course would benefit from wall-to-wall applications. The waste generated at Richard's course consists of greens clippings, thatch, leaves, harvested cores, pond dredging, kitchen wastes, and garden wastes. Twice annually, he cleans out all the drainage ditches on the course which generates ~15 yd^3 per year of wastes and once every three years the ditches are redug with a backhoe, which produces 140–210 yd^3 of waste (some of which is soil).

During the composting process, the mass of these residues can be reduced by 70% or more so 100 yd^3 of waste may only produce 30 yd^3 of compost. A ⅛ inch 50/50 (sand/compost) topdress layer, covering a 5,000 ft^2 green, will require ~1 yd^3 of compost.

Accepting wastes from outside sources is generally not recommended unless one is very familiar with the source. There is often little control over what is received and there is a greater risk of contamination. Additionally, as one enters the realm of commercial composting, special permits are almost always required. We interviewed one manager who only accepts waste from landscapers equipped with vacuum trucks. This equipment picks up only lighter material such as dry leaves and clippings but leaves glass, metal, and other contaminants behind.

In the event that the space, location, or regulatory obstacles cannot be overcome, off-site composting may be a viable alternative. Vendors who supply sand, topsoil, sod, plants, equipment, or even chemicals may have available yard space and equipment to produce quality compost. If there is some profit in the proposal for them, they may be more than willing to commit the space.

A grass pile with a free-lift stake can be found at every green on co-author Luff's course. Placement is strategic to minimize interference with the game. Piles are collected 2-4 times per year and added to the nearest compost heap.

Commercial compost producers located nearby may also be able to work with the superintendent. Most of the producers we've interviewed would be willing to cure, according to the superintendent's specifications, enough compost for the entire course. They also may be willing to take delivery of organic wastes from the golf course and use it in the piles designated for the course. Contracting with a commercial compost maker is a great way to avoid regulatory complications but only if the producer is willing to commit the time and space needed to produce a well-aged, disease suppressive compost. The compost maker would have to receive a fair price that compensates his extra time, effort, and the monopolization of some yard space but chances are that a good relationship can be established between the superintendent and the compost producer. The superintendent should be willing to make a few inspection trips to the site while the compost is curing and bioanalysis should be done before the finished product is delivered. If any biostimulants need to be added, it should be done at the manufacturer's site where the necessary equipment and space is readily available.

Ensuring a good quality, disease suppressive compost requires a certain amount of analysis (at least until one gains ample experience) to determine if it is within acceptable parameters. These tests can include measurements of organic matter, pH, conductivity, respiration, nitrate, nitrite, sulfide, ammonium, ammonia, C:N ratio, and biological activity. These tests have varying degrees of importance and some will negate the need for others. Disease suppressive characteristics should be measured by a bioanalytical laboratory where disease suppressive organisms can either be identified or actual disease suppression can be evaluated.

Respiration may be the only test an experienced compost-maker needs to determine the stability of his compost. Simple on-site testers can measure the amount of carbon dioxide being generated from compost and determine its degree of maturity. Immature compost generates much higher levels of carbon dioxide than mature compost and test kits like the Solvita® Compost Maturity Test (Woods End Research) can measure how much is being generated. Maturity, however, is not necessarily an indicator of disease suppressive characteristics. If a superintendent is relying on compost to suppress disease, it is important that a qualified laboratory performs a bioassay.

Conductivity determines the saltiness of compost and can be another indicator of maturity. Usually, values above 3 millimhos or siemens (1920 mg/kg salt) indicate a less than mature compost. Conductivity can be measured with on-site equipment (Hach Company, Hanna Instruments) or a sample of compost can be sent to an off-premise laboratory for analysis. There are factors that can influence both the respiration and conductivity tests but most of them differ for each test so both can offer confirmation of each other. The Solvita® Compost Maturity Test actually measures both carbon dioxide and ammonia simultaneously and the combined results factor into the interpretation of maturity. Woods End Research combined the two tests because a restriction of oxygen to decay organisms can reduce their ability to respire carbon dioxide which, by itself, might indicate a mature compost. A condition in compost approaching anaerobiosis, however, would produce relatively high levels of ammonia. Woods End Research recommends adequate moisture content before testing because neither carbon dioxide nor ammonia will be produced in measurable quantities if moisture is deficient.

The Cress Test is an on-site maturity test where cress seed is germinated in a flat filled with compost or in absorbent material soaked with aqueous compost extract. Maturity is based on if the seeds germinate and how much the plants grow in six days at 28° C (82.4° F). If the seeds are unable to germinate, the compost is most likely too young. Ideal growth ranges are 60–100 grams of plant matter grown per 10 grams of seed in six days; 30 grams, however, is acceptable. This is generally an infallible test but it requires a sensitive scale and six days' lead time to measure results. Some managers use grass or other types of seed. A good germination rate and plant health should be noticeable if the compost is mature. Again, maturity is not a measurement of potential disease suppression.

Analyses of nitrate, nitrite, sulfide, ammonia, or ammonium are generally unnecessary if a conductivity test has been performed. These compounds are some of the main contributors of salinity that will be detected in a conductivity test. If analyzing for these compounds, the parameters are as follows:

- nitrate – 100 to 300 mg/kg (depending on ambient temperature).
- nitrite or sulfide – zero to trace amounts.
- ammonia, ammonium – 0.2 to 2 mg/kg (3.0 mg/kg = critical limit).

The analytical instruments needed for on-site testing can be acquired from Hanna Instruments or Hach Company, among others (see Sources and Resources for contact information).

Organic matter content is important to know especially if compost is being purchased from an outside source. Standards set by the U.S. Compost Council require at least 60% organic matter, on a dry basis, for a material to be labeled compost. Some compost producers add sand, soil, or use an abundance of litter with low organic matter and high mineral content. Compost with a low level of organic matter can still be a fine soil amendment and improve plant health but may not provide adequate disease suppression.

pH is another test that may only be necessary as one is gaining experience. If a compost pile is heating properly each time it is turned, chances are that the pH is within the range needed for active compost organisms. The ideal pH for compost materials is between 6.5 and 8.5 but materials with a pH of 5.5 to 9.0 can still create a hot and active pile. If the pH of the raw ingredients is too low or too high, the growth of composting organisms may be retarded. This condition will result in a pile that fails to heat properly. It's important to remember, however, that organic residues will eventually rot despite their pH. Materials with a high pH such as manure can neutralize materials with a low pH such as pine needles and oak leaves. Lime can be added if necessary, but decay organisms can usually alter the pH to a more favorable range for composting to occur.

Nutrient testing is important because fertilizer programs usually need to be adjusted when compost has been applied. Even though the analysis of compost is low, the amount that is normally applied can supply a significant amount of nutrients. If, for example, compost containing 1% nitrogen is mixed 1:1 with sand and applied as a $\frac{3}{16}$-inch topdress, then half of 0.58 yd^3 (the total volume of the mixture per 1,000 ft^2) or 0.29 yd^3 of compost is being applied per 1,000 ft^2. If the compost weights 1,000 lb per yd^3, and 1% of the weight is nitrogen (N), then 2.9 lb of N are being applied per 1,000 ft^2. This may seem like a lot of N but it is not soluble or immediately available so plants aren't overwhelmed. However, adding almost 3 lb of N per 1,000 ft^2 to an already established fertilizer program could overdo it on the nitrogen to the point where other problems are created. Knowing the analysis of your compost could also save some money on fertilizer bills. Some of the superintendents we interviewed have replaced their normal fertilizer program with compost or compost tea. They keep the fertilizer in the shed but haven't had to use much since they began applying compost.

Bioanalysis is important if the superintendent is going to rely on his compost for disease suppression. There are many laboratories that can assay compost and either identify disease suppressive organisms within the compost or introduce the compost to specific pathogens and measure the amount of inhibition. Laboratories such as Soil Food Web, Woods End Research, and BBC Laboratories can perform a microbial profile, pathogen inhibition assay, or both. They can also perform other compost tests such as compost maturity, compost stability, and VAM (mycorrhiza) analysis (see Sources and Resources for contact information).

There are a wide variety of materials that can be added to compost that further enhances its ability to grow healthy plants and suppress disease. Coauthor Richard Luff makes compost from grass clippings, thatch, garden wastes, leaves, soil cores, pond muck, and other organic wastes from the course. He occasionally adds greensand, cottonseed meal, flower hull ash, and seaweed from nearby beaches. Other managers add materials like volcanic rock dusts, bentonite clay, alfalfa meal, basalt dust, wood ashes, or soybean meal. These additions all serve to encourage larger and more diverse populations of organisms in the compost.

Inoculants or enzymes are used occasionally to accelerate the composting process but these sometimes fail to produce the desired results because the bacteria or enzymes being applied are not matched correctly to the material in the compost pile. Indigenous decay organisms will eventually inoculate the compost pile regardless of what inoculum is added but introducing the right strains of organisms can decrease composting time and improve the balance of organisms in the final product. Using inoculants is the subject of considerable study and debate; however, some believe they are essential. Many compost producers use soil as an inoculum. A healthy soil contains a broad diversity of indigenous organisms and can also narrow the pile's carbon to nitrogen ratio (the C:N ratio of soil is ~12:1).

Some managers are concerned with introducing clay or silt onto their greens and, depending on how it is made, compost can contain some of each in varying amounts. Physical analysis can determine the percent sand, silt, or clay in the compost but it may be more revealing to test the topdress mixture rather than just the compost alone. Keep in mind, however, that the biological response in topdress mixtures amended with high-quality compost can physically alter not only its own texture but the texture of adjacent soil particles into a more porous and less compacted environment.

For those superintendents who just can't wait two to five years for high-quality, disease suppressive compost, controlled microbial composting (CMC) may be worth examining. CMC was developed by an Austrian couple, Siegfried and Uta Luebke. It is an intensely managed composting system which requires a special microbial inoculum, frequent turning, and additives such as clay loam, finished compost, and basalt rock dust. Monitoring equipment is also recommend that measures pH, oxygen, carbon dioxide, nitrate, nitrite, sulfide, and ammonium. Turning is required daily for the first week, six times in the second week, five times in the third week, four times in the fourth week, and twice weekly in the fifth and sixth weeks. Any water added to the pile cannot be chlorinated. The initial pile is built in layers beginning with dry, carbonaceous materials and alternating with nitrogenous materials such as grass clippings, manure, or food wastes. Covering the pile with a fleece type compost cover is also recommended. Although the process sounds complex, the end result is finished and disease suppressive compost in six to eight weeks. Companies marketing CMC technology and the special inoculum include Autrusa Compost Consulting, Herbert Ranch, Pike Lab Supplies, and Fresh Aire Implements. Sources of inoculum also include Josephine Porter Institute of Applied Biodynamics, Midwestern Bio-Systems, and Petrik Laboratories (see Sources and Resources for contact information).

Cold composting is another alternative to consider; however, we could not find any research confirming its ability to suppress disease. Cold composting is accomplished by incorporating fresh clippings or other green wastes directly into the soil with a rototiller or rotovator. Chipped wood and autumn leaves can be incorporated also but, without a proper balance of green waste, they will decompose slowly, temporarily immobilize nitrogen, and create fungal domination in the soil which is not ideal for turf or disease suppression. Cold composting eliminates the opportunity for odors to develop and may be appropriate in areas where adjacent neighbors might object to malodorous mistakes made by novice compost-makers. Over a period of time (depending on climate and other conditions), the soil becomes enriched with organic matter in the form of friable humus. The size of the plot needed depends on the amount of wastes that are generated and the equipment used to incorporate them into the soil. Two plots are ideal so that one can age while the other is being used. As the process matures, some of the enriched soil can be excavated, screened, and mixed with sand for topdressing. Although no research could be found that substantiates disease suppressive characteristics in this type of material, it is not unreasonable to assume that many of the organisms that would colonize such a rich habitat could suppress plant pathogens. A bioanalysis of the material could easily confirm this presumption. The special inoculum used to make CMC compost can also be used when grass clippings and other green wastes are being turned into the soil; however, inoculation and turning need to occur simultaneously because even brief exposure to UV light can kill the beneficial organisms.

One of the problems associated with cold composting is the survival of weed seeds. If annual bluegrass or other weed seeds are collected along with clippings and incorporated deep enough into the soil, they will not germinate and heat is not produced in this process that would kill the seed as it does in a conventional compost pile. Constant tillage will bring many seeds to the surface where they can germinate and be dispatched in the next tilling cycle but it is likely that some viable seeds will remain. Clippings can be piled upon perforated pipes and allowed to heat up for a few weeks before incorporation into the soil (see Figure 3.1). This initial heating may kill most if not all of the seeds that may be combined with the clippings.

Cold composting will eventually and undoubtedly involve the work of earthworms to varying degrees. Constant use of the tiller can suppress earthworm activities but it will be hard for them to resist the organic residues being added to the soil. If cold composting is impractical, vermicomposting (worm composting) is a process that may be manageable. The conditions inside an earthworm are similar to the inside of a well-made compost pile. Decay organisms flourish inside the worm and, it is believed, the worm extracts nourishment with the assistance of these organisms. Consequently, vermicompost is biologically active and disease suppressive. Unfortunately, like cold compost, vermicompost doesn't generate heat and cannot kill weed seeds. Brief, hot composting before vermicomposting, as discussed with cold composting, may be a practical alternative. Vermicompost is usually done in bins so that the earthworms do not escape into soil depths but it can be done successfully on the ground or in sunken beds. As long as resources are abundant, it is unlikely that worms will venture very far. Worms work from the bottom up so fresh material is always added to the top of a vermicompost pile or windrow. Soil makes a good bottom layer and fresh clippings, garden wastes, thatch, or food wastes, can be layered over the top. Diversity of feedstock is important but worms are unable to digest coarse woody material easily. Inoculating the pile or windrow with worms such as red worms (*Lumbricus rubellus*) or brandling worms (*Eisenia foetida*) that specialize in vermicomposting is a common practice. About one pound of worms per ft^3 of organic residues are usually added. Worms will reproduce so further inoculation is generally unnecessary. Source of worms can be found on the web by typing the word 'vermicompost' into a search engine. Some sources include yelmworms.com, happydranch.com, and vermico.com. When compost is needed, it has to be excavated from the bottom of the pile. Since most of the worms will be feeding in the top layer where the freshest residues are, the top portion can be carefully removed and reused as the inoculum for another pile or replaced onto the existing pile after the lower layers of compost have been removed. Vermicomposting should be done in a shady area without exposure to excessive wind. It is important to keep the pile moist. If the region is subjected to deep and sustained frost during the winter, bins cannot be used or the worms will probably perish. However, depending on their size, bins can be moved indoors where vermicomposting can continue through the winter.

Cold winters can freeze a compost pile if it is not actively heating or too small to generate and maintain high temperatures in a cold environment. Biologically active compost in a good-sized pile (>10 yd^3) should be able to maintain internal warmth even in very cold ambient temperatures. Anaerobic piles often freeze solid. The outer layer of curing piles may freeze depending on their stage of maturity but it is unusual for them to freeze to the core. If compost is needed in the early spring for topdressing or repairs, it is advisable to bring some indoors for the winter. It's important, however, not to let the material become excessively dry over the winter or valuable organisms may be lost.

Site selection for producing compost is an important consideration. Shady areas conserve moisture and moderate pile temperature. Windy sites that could potentially dry out

Regardless of how well or how old a compost is, it still has to be screened to use it on greens. This drum-type compost screener is commonly used by big

the pile should be avoided. A solid, well-drained pad composed of packed gravel is ideal for larger operations. Concrete pads are practical for smaller piles. In Europe, almost all farms, no matter the size, have a concrete compost pad consisting of a floor and back wall. They are usually built into a hillside so organic wastes can be easily unloaded from the bed of a truck or wagon parked on the uphill side and usually inclined downward toward the pad. Some pads have one or two side-walls. Cement pads decrease the likelihood of picking up rocks or other large objects that will need to be screened out.

Screening is a necessary step if compost is being used on a golf course, especially on greens. The longer the compost cures (within reason), the less likely it is that there will be chunks of partially decayed material. However, screening the material is still necessary. Screen size is usually a personal preference. The smaller the mesh, the finer the finished product but screening through a smaller mesh usually takes longer and produces more waste material. One-fourth inch mesh is a good compromise for compost designated for the golf course. Although time-consuming, double screening produces a more consistent material. Some superintendents will mix their compost with sand first and then screen them together and others screen just the compost and then mix it with sand. Premixing with sand often helps the material flow over the screens better but sand that clings to oversized compost particles will be screened out. The screening machines are relatively expensive but last for many years. Some tool rental businesses may have screeners available. Companies that normally mix and screen topdress materials for golf courses may be willing to provide the service. The superintendent may have to truck his compost to someone who can screen material unless the compost is being produced by the same people. Once the value of mature, well-made compost is discovered, the investment for related equipment will be much easier to justify.

There is still a lot to be discovered about the benefits of compost as compared with the benefits of pesticides, for example. Pesticide manufacturers sell nearly $12 billion

producers but smaller units are available. Screening equipment can be expensive but can also last for decades.

(yes, that's billion with a "b") worth of products annually in the United States alone and, accordingly, there is a significant amount of funding available for research (especially research that makes these products look attractive). The compost industry, by comparison, is growing but has only a small amount of resources available for research. Most of the available funds are invested in solving solid waste disposal problems, not dollar spot or brown patch problems. Compost is used more widely in agriculture than in horticulture and it follows that research done from the perspective of pedologic ecosystems will address crop concerns more than ornamental, or more specifically, golf course concerns. Consequently, there are no real standards by which compost is made for the golf course. Many different manufacturers produce compost from many different organic residues but relatively few produce it to the degree that is discussed in this chapter. In fact, it would be nearly impossible to find two different composts, regardless of the quality, with exactly the same chemical, physical, and biological characteristics. The quality of compost that superintendents need, however, is not the commodity that is commonly being marketed.

Although the intent of the chapter is to familiarize the superintendent with the benefits and mechanisms of composting, the information is, without question, incomplete. Even if we included all that is known about compost—which would take thousands of pages—it would still be incomplete. Given the almost infinite number of variables associated with all the aspects of creating compost, what has been discovered thus far could easily be only the tip of the iceberg. The task and the rewards of discovery are, to a certain extent, left to the ingenuity of the superintendent.

Chapter 4:

Analysis

Many turf managers tend to apply lime and fertilizers by instinct. They have developed a program that, over time, has become routine. This system can work for years, perhaps decades, but after awhile, strange little problems may arise—problems that can eventually grow into bigger ones. What has happened? While we weren't looking, our program may have slowly changed the balance of fertility in the soil. Turf utilization of a particular nutrient may have exceeded the supply, or the buildup of one nutrient may have caused the deficiency of another. Some superintendents rely on fertilizer manufacturers for recommendations. Manufacturers usually have multistep programs that make the manager's life easier. However, we trust that the directions on the bag, box, or bottle of fertilizing material we are about to apply take into consideration all of the vagaries and idiosyncrasies of our soil. They don't. Standard fertilizer recommendations are made based on several broad assumptions. The first is that climate (temperature, precipitation, etc.) will be typical at every location and that all species of turf have the same nutritional requirements. They also assume that all soil types are basically the same and last but not least, that the existing fertility is already balanced. These programs are not meant to correct imbalances. How could they? The fertilizer manufacturer has no idea of the soil conditions on every golf course. The only accurate way to determine those conditions is to perform a soil test.

Some of us fail to understand that the soil is not just an inert medium where turf can be grown hydroponically. It is an ecosystem that functions in a symbiotic relationship with all plants. The best tools we have to follow the changes that occur in this system are soil, plant tissue, and irrigation water analyses. Leaf tissue analysis reveals the level of nutrients being absorbed into the plant. A nutrient may appear as adequate or optimal in a soil test but plants may not be absorbing it. Water analysis is important for any superintendent who uses irrigation. Many water supplies have nutrients, toxins, or inappropriate pH that adversely affect the growth or health of turf. Soil analysis tells us the chemical balance of nutrients (or toxins), the balance of soil organisms, and the physical structure of the soil. Unfortunately, the inherent weakness of analysis is extracting a sample that represents overall or average conditions.

Imagine if you will Times Square in New York City on New Year's Eve. There are wall-to-wall people and every available place to stand is occupied. Now imagine an alien from another galaxy on a scientific expedition to gather information about humans. From his space vehicle he focuses his scanner on someone and his computer records important statistics such as height, weight, limb length and thickness, hair color, skin color, clothing, voice pattern, fingerprints, brainwaves, cell analysis, etc. The alien has his data and prepares for departure when at the last minute he decides to take another scan from a different person to make sure his equipment is working accurately. To his utter dismay, the data from the next scan is so completely different that he is unable to determine anything specific about humans. He checks his equipment and performs yet another scan. Again, there are wild fluctuations in the data. He does several more scans and cannot find any two subjects that are alike. After more than a hundred scans, all significantly different, the alien leaves in frustration. Eventually, the alien understands that the biological differences in humans are infinite and the best analysis he can make will come from averaging his data and determining typical statistics. After doing so, he can make some general observations—humans like to stand outside in cold weather, drink beverages containing alcohol, make noise, and kiss at the stroke of midnight.

It would be rare to find two soil samples that produce the same test results, even if they were drawn a foot away from each other. The only way to effectively evaluate the soil in a given area is to draw many samples and mix them together. Then a single sample representing the characteristics of all samples can be drawn from the mixture. Random sampling can be done in a zigzag or a matrix pattern. The number of samples taken should depend on the size of the area. Think of sampling this way: If you were in a room with two other people who represented the population of the United States and, after asking how many smokers there were in the room, one raised his hand, the logical assumption would be that 50% of the population smokes. If, however, there were 10,000 people in the room you would get a better representation of the U.S. population. We're not suggesting that 10,000 samples be drawn but the more samples taken, the better the representation. Drawing samples from notably good or bad areas should be avoided (unless evaluation of a specific spot is needed). The conditions in these areas are extremes of one sort or another and can adulterate the average reading of an area. For obvious reasons, freshly fertilized or limed areas should also be avoided.

Very clean tools should be used for gathering soil samples. A small amount of rust on a shovel might indicate a promising location to start an iron mining operation. An example of an incident that gave misleading results from contaminated tools is the client who used the same shovel to draw samples that he normally uses to remove ash from his wood stove. The results of the test were so far off that they might as well have been from another planet. It's unlikely that any superintendent would ever use a shovel to draw samples. Stainless steel or chrome plated sampling tubes are the most efficient and accurate tools for drawing soil. They significantly increase the speed of the sampling procedure while decreasing labor costs and the chances of contaminating the sample. These tubes are durable, relatively inexpensive, and can be purchased from most horticultural suppliers. Even if sampling occurs only once a year, these tubes are worth the investment. Using one of these tools on a green will leave core holes that can disrupt the game but sampling can be done in conjunction with a routine coring operation. In fact, if the superintendent is extracting cores that are four to five inches long, several of them can be gathered to represent the conditions of the green. Most managers don't usually core that deeply but the holes left from drawing cores with a sampling tube will be filled with topdress material after cultivation. If analysis is routinely done annually, biannually, or at other intervals, samples should be drawn during the same part of the season. The changes that naturally occur in the soil between spring and fall can give a false impression of progress (or the lack of it).

Wet or frozen samples are difficult to handle and test. The consistency of a soil sample for analysis should be moist but not soaked. Water should not drip from the sample when squeezed. Drying out the sample is an acceptable procedure if necessary, but to preserve the original conditions of the soil, samples should not be drawn until the soil is at a proper moisture level. If the soil in question is naturally sandy and dry, it is not necessary or advisable to moisten it.

The depth from which a soil sample should be drawn is normally about four to five inches for turf. The fertility of most soils increases nearer to the surface. Deeper samples tend to underrepresent the fertility of the root zone and shallower sampling will often overrepresent it. If testing is routinely done for comparison purposes, it is important that samples be drawn from the same depth each time. Surface debris such as sod, roots, or thatch should be removed from the top of the sample. When enough samples from a given area are drawn, they should be mixed thoroughly in a clean container and a representative sample of approximately one-cup can then be sent to a lab. The soil is a relatively dynamic substance and, if the sample is allowed to sit for an extended period of time, subtle changes may occur that, when magnified by the area it represents, may yield misleading results. Send it to the lab as soon as possible.

Analytical laboratories are not obligated to use nationwide standard procedures for testing soil. Many labs in a given area may often employ regional methodologies, but they can opt to follow any procedures they want. *Methods of Soil Analysis*, published by the American Society of Agronomy, is the most widely accepted manual of testing procedures; still, the publication often describes many different yet valid ways to test for nutrients and other soil conditions. If several different labs analyzed the same sample, each using a different methodology, the results may appear as though the samples were taken from several different regions of the country. Additionally, the standard set of tests one lab does may be significantly different than another. Some labs, for example, calculate the cation exchange capacity (CEC) of the sample while others calculate an effective cation exchange capacity (ECEC). The difference between the two (discussed later in this chapter) is significant. Some labs base lime recommendations on a buffer pH test while others on a test for exchangeable aluminum. Because analyses from different labs often tender significantly different results, it is a good idea to choose a reliable lab and stick with it, especially if one is comparing tests with preceding or subsequent analyses. On a functional level, the most important service a lab can offer is an accurate, reasonably priced test delivered on a timely basis. It is also important to look at the kind of information offered by different labs. It is most cost-effective if all the necessary information is offered on the lab's standard test, since optional information can get expensive.

The American Association of Laboratory Accreditation (see A2LA in Sources and Resources for contact information) is an organization that accredits testing labs in many different fields. The United States Golf Association (USGA) recognizes eight of these labs as competent to analyze materials specified by them for putting greens; however, there are other accredited labs that may offer more specialized information for ecological management (see www.A2LA.org for more information). The USGA list includes Brookside Laboratories, Inc.; European Turfgrass Laboratories Limited; ISTRC New Mix Lab LLC; Links Analytical; N.W. Hummel & Co.; Thomas Turf Services, Inc.; Tifton Physical Soil Testing Laboratory, Inc.; and Turf Diagnostics & Design, Inc. (contact information can be found in the Sources and Resources section.) These labs specialize in physical soil analysis for turf environments but Brookside is also equipped to perform nutrient analysis as well. Most of these labs are associated with other analytical companies so if they can't offer a specific analysis, they can have it done elsewhere. Some may offer biological analysis as well.

Soil test interpretations are sometimes as different as the methodologies used to determine the results. If the superintendent uses the same lab each time analysis is done, interpretation should be relatively easy to learn. Unfortunately, some labs give recommendations, along with insufficient data to make one's own accurate interpretation. Those labs generally cater to customers who only want the information necessary to balance and fertilize; they are not really interested in learning what the information means. Understanding how soil chemicals are measured and how they react with one another enables the superintendent to consider variables—unique to his course—about which the lab technician couldn't possibly know.

Lab recommendations are usually offered automatically but sometimes as an option (extra cost). These recommendations can only be correct if the samples accurately represent the average conditions of the area in question. Recommendations are generally based on the nutrient needs of specific plants under average conditions but some labs will take the soil's ability to store and exchange nutrients into consideration. Some labs will ask for more information such as type of grass, topography, acreage, and previous treatments before providing recommendations.

Managers who wish to implement ecological or organic practices are more or less forced to interpret the analysis reports themselves. Labs that offer nonchemical recommendations are few and far between. The main obstacle for most labs is that chemical analysis cannot fully reveal biological conditions. Nutrient deficiencies and imbalances often affect plant growth but can also influence other soil organisms as well. An optimum balance of nutrients, however, doesn't always guarantee adequate plant absorption. There are numerous instances where adequate nutrients are present in the soil but leaf analysis reveals deficiencies in the plant. This can occur when soil chemical reactions bind certain nutrients into unavailable forms and biological activity is insufficient to free them. There are also instances where soil organisms will immobilize plant nutrients. Biological analysis is not yet considered in vogue but more and more managers are recognizing its value in evaluating soil conditions.

This is not to say that traditional soil chemical analysis is not important. The condition of each soil component (i.e., physical, chemical, biological, etc.) affects all others. To better organize this information, we will exhaust what we can share about chemical analysis before moving on to physical, biological, water, or leaf tissue analysis.

Standard information on a soil analysis report can vary from lab to lab but most will provide values for pH (water and buffer), cation exchange capacity (CEC), reserve and available phosphorus (or phosphate), levels of exchangeable potassium, magnesium, and calcium, and base saturation. Many labs also determine the percent of organic matter, which is a valuable measurement. Most labs offer nitrogen or micronutrient analysis as an option.

pH is probably one of the most important tests that can be performed for turf. It can affect the availability of most nutrients (and toxins), the soil's structure, and the balance of soil organisms. Some believe that the initials pH stand for potential hydrogen or power of hydrogen. The H in pH does, in fact, stand for hydrogen but p is a mathematical expression which, when multiplied by the concentration of hydrogen in the soil, gives a value between one and fourteen that expresses the acidity or alkalinity of the soil. Actually, soil pH values are rarely found to be less than 4.0 or over 10.0. Values below 7.0 are acidic and those above 7.0 are alkaline. A value of 7.0 is neutral. As values get farther away from neutral they indicate a stronger acid or base. A soil pH between 6.0

and 7.0 is ideal for most plants. Most species of turf plants grow best at a pH of between 6.5 and 6.9.

Almost all labs offer a pH measurement in their soil analyses. Most labs conduct a water pH analysis, which determines the acidity or alkalinity of distilled water when mixed with an equal volume of soil. A few labs will perform a salt pH analysis that takes into consideration the seasonal variation of soil soluble salts that can cause changes in pH. Salt pH values are normally 0.5 to 0.6 below what a water pH test would show. These tests determine whether or not there is a need to lime or acidify the soil. Acidification recommendations can be made from the results of a water or salt pH analysis but usually, lime recommendations are not. Another test is generally needed to determine lime applications because the amount of reactive hydrogen held in the soil may vary significantly. Two different soils with the same water pH can have completely different lime recommendations. A sandy soil, for example, with a pH of 5.5 may require only a half-ton of lime or less per acre to raise the pH to 6.5 whereas a clay soil with the same pH may need over four tons of lime per acre to achieve the same result. There are several acceptable methods of determining lime requirements. These include a buffer pH test, reactive aluminum test, titration of soil acidity, or buffer estimations based on soil organic matter, soil texture, or both.

The buffer pH test is done with a special solution that determines the soil's capacity to hold exchangeable hydrogen (H^+) or hydronium ions (H_3O^+). The buffer test is only used if the water pH indicates a need for lime. If only one pH value is shown in an analysis report and there is no indication of how the test was performed, it is relatively safe to assume it is a water pH. The lab's recommendation for lime, however, is usually based on something more than a water pH test and it would be a good idea to contact them and ask. Chances are good that the recommendation is based on the amount of organic matter in or the texture of the sample. Some labs will test for exchangeable aluminum and, through a complex logarithmic formula, offer liming recommendations. This method was originally developed for semitropical and tropical regions where toxic aluminum is the most limiting factor to crop production. The titration method measures the actual amount of calcium carbonate equivalent (CCE) needed to neutralize the sample. Carbonate is the component in lime (calcium carbonate or calcium/magnesium carbonate) that neutralizes acidity in the soil.

The most commonly used methodology of the four mentioned above is the buffer test. There are two distinctly different buffer tests commonly used, the SMP (Shoemaker-McLean-Pratt) and the Adams-Evans buffer. The SMP test is used mainly in the Northeast where the soils commonly have a significant reserve of exchangeable aluminum. The Adams-Evans buffer is more appropriate for sandy soil that has a low CEC and a low level of organic matter. Golf course greens that contain a lot of sand might be better served with lime recommendations based on the Adams-Evans buffer but some analysts believe all buffer tests are too insensitive when it comes to extremely sandy soil. Greens that contain calcareous sand, however, may not need lime at all. This would be apparent from the water pH test.

One of the reasons that different soils with the same pH have different lime requirements is that their capacity to hold cations, such as H^+, K^+, Ca^{++}, Mg^{++}, and NH_4^+, varies. To better understand cations (positively charged ions) and anions (negatively charged ions), a short review of elementary chemistry is needed. All materials that are in, on, or above the planet, whether liquid, solid, or gaseous, are comprised of atoms from different elements that are often bonded to each other. The force that holds these atoms together is electromagnetic. It is similar to the force that holds dust to a computer screen. In order for this magnetic force to work, however, there must be both positively and negatively charged ions present. Like charges do not attract.

There are 90 naturally occurring elements on earth. Elements are made up of atoms and atoms are a combination of neutrons, protons, and electrons. The electrons move around the neutrons and protons (bunched in the center of the atom) in orbit paths that differ in distance from the nucleus. Each path holds a set amount of electrons. If the outermost path is filled to its capacity then it has no charge and rarely bonds with anything. These elements are called inert. If there are vacancies in the outermost electron path, then the atom can share electrons with other atoms, creating a compound. These vacancies give atoms either a positive or negative charge depending on whether they can take on or give up electrons. Most elements are made of atoms that have either a positive or negative charge. When the atom of one element bonds with the atom of another, the result is a molecule of a compound. The magnetic bonds between elements in many compounds satisfy the attraction each atom has for the other. These compounds are left with neither a positive nor a negative charge. If, however, the electrons in the outer ring of one atom do not completely fill the outer ring of another, a net negative or net positive charge is still available for yet another combination. These compounds are called ionic.

Ions can be atoms of elements or molecules of compounds but they both carry either a negative or positive electromagnetic charge. The ions of elements such as hydrogen (H), calcium (Ca), magnesium (Mg), and potassium (K) have positive charges and are known as cations. Ions of phosphorus (P), nitrogen (N), and sulfur (S) have negative charges and are called anions. Some elements such as carbon (C) and silicon (Si) can act as anions or cations and bond to either charge. Ionic molecules such as nitrate (NO_3), sulfate (SO_4), and phosphate (PO_4) have negative charges and can bond with cations such as H, Ca, or K.

There are very small particles in the soil, called micelles (short for microcells) that carry an electronegative charge. These particles are either mineral (clay) or organic (humus) and are referred to as soil colloids. Cations are attracted to these colloids like dust is to a TV screen.

Clay particles, viewed through a powerful microscope, appear as flat platelets adhering to each other like wet panes of glass. These particles are predominately comprised of silicon, aluminum, and oxygen but, depending on the type of clay, can contain a plethora of different elements such as potassium (K), magnesium (Mg), iron (Fe), copper (Cu), and zinc (Zn). Clay's magnetism comes from the natural substitution of ions in its structure with other ions that don't completely satisfy the available magnetic charges. As more of these substitutions occur, the overall negative charge of the particle increases, and there is a relative increase in the amount of cations attracted to it (see Figure 1.2). Different types of clays have different abilities for ionic substitutions, and their magnetic force differs accordingly. Clay particles that have greater substitution potential will create more magnetic force and attract and hold more cations. Some clay particles are not colloidal at all. Clay is not generally found in abundance on golf courses, especially on greens and, even though there is evidence that a small amount of clay can significantly improve soil conditions, we are not suggesting indiscriminate use of clay as a soil amendment to improve CEC.

Humus particles are also colloidal. Their cation exchange capacity per equal unit of weight is even greater than that of clay but it is influenced by soil pH (see Figure 4.1). Humus becomes magnetically charged from the surface compounds that contain hydro-

FIGURE 4.1. INFLUENCE OF pH ON THE CEC OF ORGANIC MATTER AND CLAY

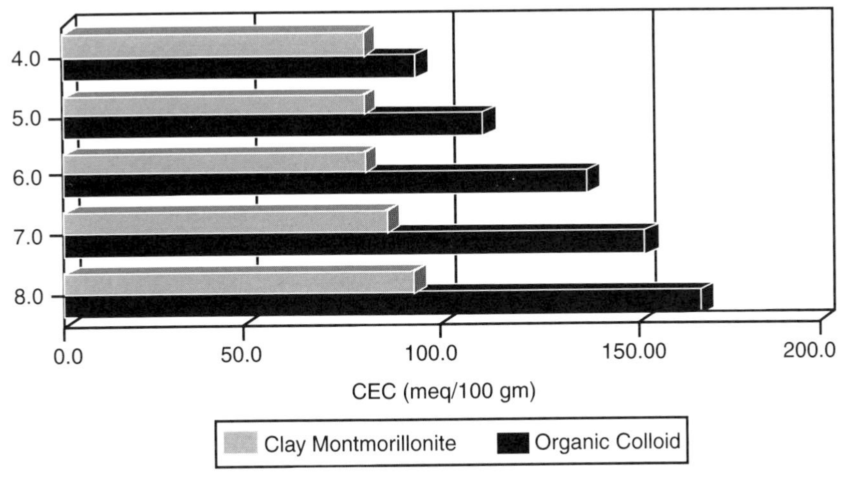

Adapted from Waksman, 1936.

gen (H). In soils with a near neutral pH, H is displaced from humus to participate in a number of different soil chemical reactions. When this occurs, the site that previously held the H ion is left vacant and available for another positively charged ion. If base cations such as K, Mg, or Ca are present in the soil solution they can be attracted to the humic particle (see Figure 4.2).

Humus and clay can form colloidal complexes together, which not only increases the overall cation holding capacity of a soil, but also changes the structure of clay soil into a better habitat for turf roots and aerobic soil organisms. Humus, through complicated

FIGURE 4.2. ORGANIC COLLOID

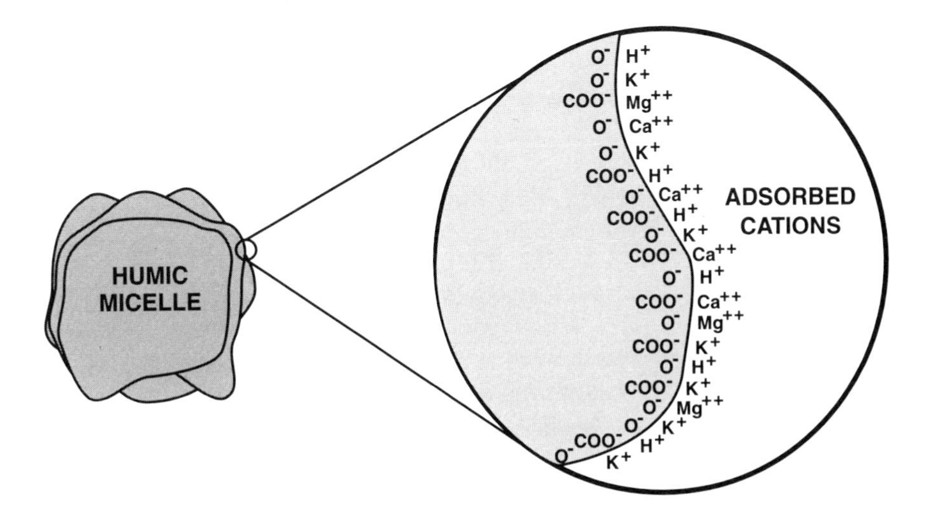

FIGURE 4.3. CATION EXCHANGE

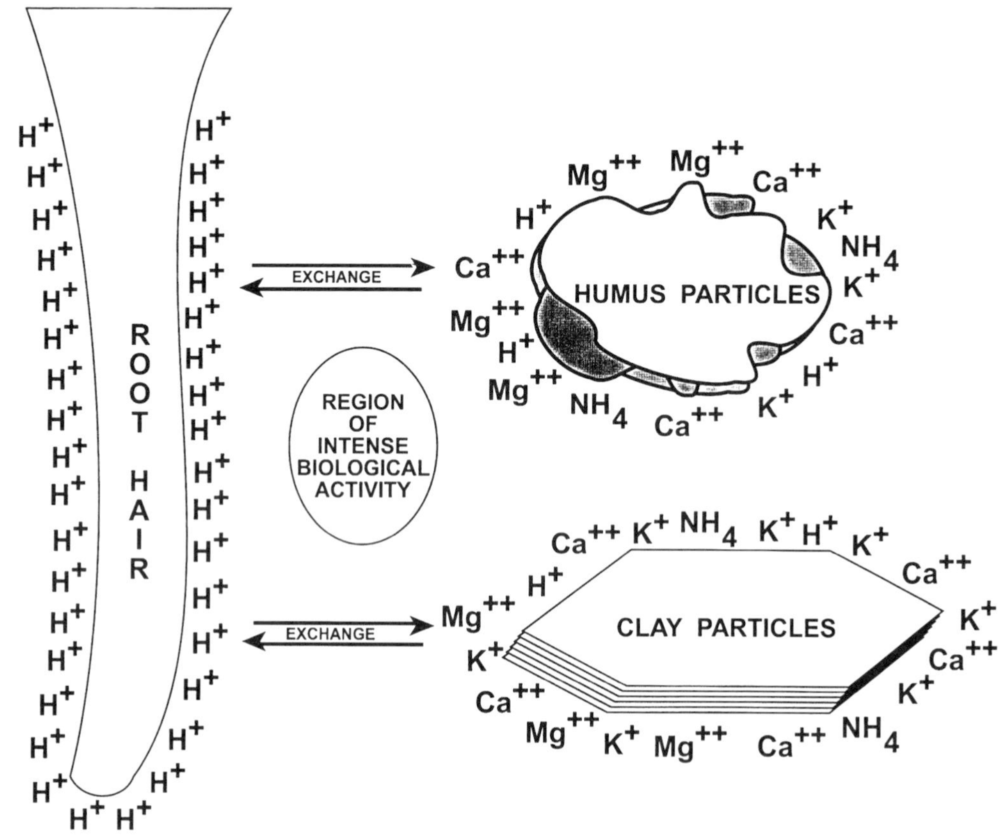

chemical bonds, can surround clay particles, breaking up clay's cohesive nature that often prevents percolation of air and water through the soil. These bonds can also help extend the length of time humus particles exist in the soil.

Both clay and humus, with their unique structure and negative charge, adsorb (as opposed to absorb) cation elements in such a way that the ions can be detached and absorbed by plant roots. Adsorption is simply an adhesion of ions to a particle surface whereas absorption is more like consumption or assimilation. Hydrogen (H^+) ions that are released by roots are traded back to the colloids. This process of cation exchange is nature's way of providing a sustainable source of cation nutrients for plants (see Figure 4.3). Essential nutrients such as calcium, magnesium, and potassium would be rare and unavailable in topsoil if it were not for the colloidal nature of clay and humus. When root hairs grow into proximity with exchange sites, cation nutrients from the colloid can displace hydrogen ions located on the root's surface and be absorbed by the plant. Cation exchange is essential, not only for the mineral needs of the plant, but also for the translocation of hydrogen throughout the soil.

Most soils are mineral soils because they were formed from rocks weathered into smaller and smaller particles by forces such as rain, frost, wind, and erosion. Eventually, the surface area of these mineral particles increased to a point where certain organisms could access some of its mineral nutrients. These pioneers of soil formation were likely photosynthesizing organisms and their needs consisted mostly of gaseous elements such as oxygen, nitrogen, hydrogen, and carbon dioxide derived from the atmosphere. Some of what they needed, however, was attached to the weathered rock particles that surrounded them. As generations of these photosynthesizing organisms cycled through life and death, decay and predacious organisms began to evolve. As time passed, the accumulation of organic residues from the biomass served to support a greater diversity and even larger populations of organisms and eventually higher plants. Contributions of residues from plants accelerated the growth and diversity of the biomass and the creation of humus. Organic matter has its own life cycle, however, and through oxidation, nitrification, and other natural processes would, on average, amount to approximately 5% in the top six inches of a mineral soil in the temperate zone of the planet. This figure varies as climatic and other conditions change from location to location.

Organic matter is often a barometer of soil health and measuring it is a useful practice. The population of organisms supported by soil organic matter is of immeasurable benefit to plants. More organic matter usually means:

- more decomposers and faster recycling of nutrients (including carbon dioxide) from plant and animal residues;
- more nitrogen fixing, phosphorus solubilizing, and hormone producing soil organisms;
- more beneficial organisms that help dissolve mineral, transport water from soil depths, and antagonize plant pathogens;
- more soil-dwelling predators that release plant nutrients; and
- more humus to increase the water and nutrient holding capacity of the soil.

Humus expands and contracts in the soil as its moisture level changes. This phenomenon increases porosity, which improves the movement of water and the exchange of gases with the atmosphere and alleviates compaction.

It is important when drawing samples for organic matter analysis that any undecomposed residues such as crowns, roots, or thatch be carefully removed from the sample. These residues do not represent the soil organic matter fraction and will distort the results of the analysis.

Burning is a standard method of testing for organic matter. The sample is dried before it is burned to minimize water weight that will inevitably be calculated as part of the organic carbon driven off by combustion. Analytical chemists admit that even after thorough drying, some water molecules, steadfastly attached to hydrated compounds, may still remain but that the level of error from these persistent units of water is well within an acceptable range. The mineral portion of the sample doesn't burn and will remain after combustion. The percent of organic matter can then be determined simply by subtracting the weight of the ash from the total weight of the dry sample before it was burned.

The USGA recommends between 0.5 and 2% organic matter for green construction and they recognize that there will be a natural increase in that percentage over time. Generally, when the USGA refers to organic matter as excessive, they include undecomposed thatch, roots, and other residues in their measurements and, in general, are not in favor of high levels of organic matter on greens. We aren't either. If most of the organic matter in a soil is undecomposed residues it's not helping the turf, the game, or the soil organisms. Our opinion of organic matter is entirely different, however, when the bulk of it is stable or friable humus. Many would argue that an abundance of sand and lack of organic matter is necessary for a hard, well-drained playing surface. We don't argue with that logic but would ask instead: Do we need to sacrifice so many of the benefits in a biologically rich soil for excessive drainage and inordinately hard surfaces?

It may be difficult for most superintendents to maintain a level of soil organic matter (disregarding thatch and other undecomposed residues) above 5% no matter how hard they try. A biologically active soil consumes most organic residues and generates carbon dioxide for plants. Additionally, the longer and warmer the growing season, the lower the natural levels of soil organic matter will be. A reading close to 4% in temperate regions is a good level but may be difficult to attain and maintain. An analysis of soil extracted from coauthor R. Luff's greens reveals organic matter levels at between 3 and 4%. This is a surprisingly low figure after almost four decades of topdressing with compost twice annually but it represents an extremely functional soil ecosystem that consumes the energy in the compost to perform essential biological activities. Greens where soil organic matter is below 2% may not be able to maintain functional levels of biological activity. In some cases, an analysis may reveal high organic matter but may not correspond to a relative level of biological activity. Bioanalysis and respiration tests will be discussed later in this chapter. Some labs will calculate an *estimated nitrogen release* (ENR) from the amount of organic matter found in the sample. This value is usually given in pounds per acre. (To convert this value to pounds per 1,000 ft^2, multiply by 0.023.)

Many labs use a colorimeter method for determining percent organic matter in the soil. The soil is mixed with a chemical that changes color when it reacts with organic matter and the intensity of color is measured. This method is a valid and accurate way of determining the percent organic matter. Most labs will use either method if a specific request is made. Many labs routinely perform organic matter analysis but not so much because they recognize the value of humus but because the biological activity inherent in a rich soil interferes with the effectiveness of many herbicides.

Depending on soil pH and the level of exchangeable cation bases (i.e., Ca, Mg, and K) in the soil solution, the hydrogen that is adsorbed to the soil colloid may be pushed off. Once the H^+ ion enters the soil solution a number of reactions can occur. It can combine with anions to form water, gases, or acids. Acids can work to liberate more base cations from the mineral components in the soil. H^+ can also liberate and transform phosphate ions into the form of phosphorus that plants can absorb.

Soil carbonates react with H^+ ions to form carbon dioxide (CO_2) and water. This process reduces hydrogen ions in the soil and raises the pH. This is how lime (calcium carbonate) neutralizes soil acidity. Oxides, abundant components of many minerals, can also neutralize H^+ forming water (H_2O). Hydrogen can also contribute to the formation of organic acids when reacting with residues of plants, animals, and microorganisms. These acids can often weather unavailable minerals attached to parent particles into available nutrients. This function is especially important in the rhizosphere, the zone immediately surrounding roots. Within the rhizosphere, biological activity can be very high and the pH, relatively low. The activity of H^+ ions in this zone can release mineral ions from soil particles that can then become available nutrients for the plant.

The cation exchange capacity (CEC) of the soil is determined by the amount of clay and humus in the soil. However, not all clay or humus is colloidal. Certain types of clay have a much higher CEC than others. Some clay is not colloidal at all. Friable humus has a relatively low CEC compared with stable humus. The right kind of clay and stable humus are essentially the soil's cation warehouse or reservoir. Sandy soils with little organic matter have a low CEC, but clay soils, especially those that also contain high levels of organic matter, have a much greater capacity to hold cations.

The disadvantages of a low CEC include limited availability of mineral nutrients to plants and the soil's inefficient ability to hold applied nutrients. Plants can exhaust a fair amount of energy—that might otherwise be used for growth, adapting to environmental stress, and developing adequate defenses—scrounging the soil for mineral nutrients. Soluble mineral salts (e.g., potassium sulfate) applied in relatively large doses to soil with a low CEC cannot be held efficiently because the cation reservoir is too small. Plants are faced with a use-it-or-lose-it situation and they don't have the ability to store mineral nutrients.

Water also has a relationship with colloidal particles. Functions that are dependent on soil moisture are also limited in soils with low CEC. Organisms such as plants and microbes that depend upon each other's biological functions for survival are inhibited by the lack of water. A heavy clay soil with a low level of organic matter might have an opposite effect (a lack of air), causing problems associated with anaerobiosis. The CEC in such a soil might be very high, but the lack of atmosphere in the soil can limit the amount and type of organisms in the soil, causing dramatic changes to that immediate environment. A heavily compacted soil would have similar problems. Oxidized nutrients such as nitrates (NO_3) and sulfates (SO_4) may be reduced (i.e., oxygen is removed) by bacteria that need the oxygen to live, and the nitrogen and sulfur could be lost as available plant nutrients.

The CEC of a soil is a value given on a soil analysis report to indicate its capacity to hold cation nutrients. CEC is not something that is easily adjusted, however. It is a value that indicates a condition or possibly a restriction that must be considered when working with that particular soil. Unfortunately, CEC is not a packaged product. The two main types of colloidal particles in the soil are clay and humus and neither is practical to apply in large quantities. Compost, which is an excellent soil amendment, is not stable humus but over time compost can contribute to the bank of stable soil humus. This contribution, however, may only amount to 1–10% (in some cases, less) of the initial application.

Table 4.1 shows how much compost is needed at different thickness levels on a 1,000 square foot area. Each percent of organic matter found in a soil analysis is equivalent to over 450 pounds per 1,000 ft^2 (~6 inches deep) area (20,000 lb/acre). Compost normally contains about 40 to 50% organic matter on a dry basis, and weighs approximately 1,000–1,500 pounds per cubic yard (depending on how much moisture it contains). If the moisture level were 50%, it would take two yd^3 of compost per 1,000 ft^2 to raise the soil organic matter level 1% and this change would be ephemeral. Large applications of compost to the surface of the soil, however, can do more harm than good. Abrupt changes in soil layers can inhibit the movement of water and restrict the soil's capacity to hold moisture and exchange gases. Building soil organic matter is not something that can or should be done overnight.

A soil with a very low CEC can and should be adjusted but not solely because of the CEC. A soil with a very low CEC contains very little clay or

Table 4.1. Compost Spread Rates

Thickness, In.	To Cover 1,000 ft^2
1/16	0.2 yd^3
1/8	0.4 yd^3
3/16	0.6 yd^3
1/4	0.8 yd^3
5/16	1.0 yd^3
3/8	1.2 yd^3
7/16	1.4 yd^3
1/2	1.5 yd^3

humus. Its texture is probably closer to sand or gravel than to soil. It cannot hold much water or dissolved nutrients and without relatively frequent irrigation and fertilization, plants may not grow well. The adjustment is not so much to increase the CEC but to condition the soil and make it a more appropriate environment for plants. A direct result of this adjustment will eventually be a higher CEC.

During the process of soil building, the superintendent must be aware of the soil's limitations. Soil with a low CEC cannot hold many nutrients, so smaller amounts of fertilizer need to be applied more frequently. Turf growing on soil with a low CEC has nutritional habits like most infants. It can't eat a lot but must be fed often. As the CEC of the soil improves, larger doses of fertilizers can be applied less frequently.

CEC is measured in milliequivalents (meq) per one hundred grams of soil. An equivalent is actually a chemical comparison. It is a measurement of how many grams of a cationic element or compound it takes to displace one gram of hydrogen (H). A meq is simply one-thousandth of an equivalent.

This sounds complex but let's look at it in simpler terms. Picture a seat on a train. The capacity of that seat is one person. If Peter, who weighs 240 pounds, is polite enough to give up the seat to Jessica, who weighs 120 pounds, the capacity of the seat hasn't changed, just the weight of the passenger occupying it. When considering milliequivalents, think of each ion as a passenger on the soil colloid. Unlike commuters, ions are far too small to count so their numbers must be calculated by using known factors such as atomic weight and electromagnetic charge (valence). Hydrogen has an atomic weight of one and has a valence of $^{+}1$ (hydrogen is a monovalent ion). Potassium (K^+) has an atomic weight of 39 and is also monovalent. H and K are both cations and they would not normally combine, so the equivalence here is in terms of displacement or competition. It takes 39 units (weight) of K to replace one unit of H. Ca^{++} and Mg^{++} are divalent ions (i.e., they have a valence of $^{+}2$) so each would occupy two seats on the colloidal express instead of one, and their atomic weight is divided evenly between the number of seats. Therefore, it takes half their atomic weight to displace one unit of H. Ca, which has an atomic weight of 40, would need 20 units to displace one unit of H. Mg, which has an atomic weight of 24, would need 12 units.

So what has all this to do with the rate of meq per 100 grams of soil? A CEC with a value of 1 meq/100 gr means that each 100 grams of soil can magnetically hold either one milligram (mg) of H, 39 mg of K, 20 mg of Ca, or 12 mg of Mg (more likely some combination of all four). Table 4.2 shows equivalent values of common soil cations.

The CEC is a calculated measurement from the amount of exchangeable base cations (i.e., Ca, Mg, and K) and meq of H found during soil analysis. In a test where exchangeable cations are measured in parts per million, the equivalent value of the cation is multiplied by 10 and then divided into the parts per million value found in the analysis. This formula comes from the ratio between parts per million and milliequivalents per

Table 4.2. Equivalence Relation

Cation	Atomic Wt.	Valence	Equiv.	Meq. Factor
Hydrogen	1	1	1	10
Potassium	39	1	39	390
Magnesium	24	2	12	120
Calcium	40	2	20	200
Ammonium	18	1	18	180

Figure 4.4. CEC Calculation.

Base	PPM Found	Meq/100 gr
Potassium	74	0.2
Magnesium	115	1.0
Calcium	850	4.3
	Total Bases	5.5
	Hydrogen	1.0
	CEC	**6.5**

100 grams, which is ten to one. If, for example, 125 parts per million potassium were found in a soil analysis, its equivalence in meq/100 gr would be, 125 ÷ (39 × 10) ≈ 0.3. The amount of exchangeable K found in the analysis is 125 parts per million and 39 is the equivalent value of potassium that is multiplied by 10 to become the meq factor.

Once the parts per million of Ca, Mg, and K are calculated into meq/100 gr, they are added together. This total constitutes the portion of the soil's CEC currently occupied by the base cations (K, Mg, and Ca). The meq of H is calculated from a buffer pH test and then it is added to the bases to arrive at the soil's overall CEC (see Figure 4.4). An inherent weakness in this method is that some extraction methods may dissolve nonexchangeable cations. Strong acidic reactants may extract much more calcium from calcareous sand, for example, than what would normally be dissolved in the soil solution or adsorbed to colloids. The resulting CEC calculation would be misleading because most of the calcium found in this analysis is not exchangeable. Consequently, base saturation values and the Ca:Mg ratio would also be distorted.

Labs that offer a CEC value would calculate it based on their findings. If a lab does not run a buffer pH test, however, it cannot include hydrogen in its CEC calculations. Some labs offer what is called effective CEC (ECEC) which is only calculated from the meq of exchangeable base cations (K, Mg, and Ca). If a buffer pH test is not done, the lab has no H^+ value to plug into this formula. If calculating the CEC is necessary, the lab's lime recommendation can act as a de facto H^+ value. If the lab recommends three tons (6,000 lb) of lime per acre, for example, each 1,000 lb of calcium carbonate or equivalent will neutralize 1 meq of H (some analytical chemists believe the factor should be higher, i.e., 1,200 – 1,400 lb). Therefore, three tons will neutralize ~6 meq of H. According to one lab that offers ECEC, the CEC of a soil is pH dependent and will fluctuate with variances in pH. They feel that the ECEC is a more accurate measure of cation exchange capacity. Others disagree. The inherent problem with calculating the meq of H from a lime recommendation is that the recommendation may be a generalization and relatively inaccurate.

Different labs use different procedures to extract calcium, magnesium, and potassium and they may report them differently too. For example, K may be reported as K_2O (potash), or all three cation elements may be reported in pounds per acre instead of parts per million. Albeit rare, some labs do not indicate the unit of measure. It is important not to assume the relationship of these values. An erroneous assumption could easily create a disaster. It's a good idea to call the lab and find out for sure. The different extraction chemicals used by labs can paint a bleaker or rosier picture than what plant roots experience. Labs attempt to mimic the natural soil chemicals that release plant available nutrients but can't possible know the conditions in every soil. Most extraction chemicals are designed to imitate average soil conditions, which are relatively rare on most golf courses. The superintendent may need to communicate with the lab technician to ensure that he is fully aware of the course's conditions. Often, a physical analysis (discussed later in this chapter) is a good prerequisite to chemical analysis.

The balance of calcium, magnesium, and potassium in the soil can easily be calculated by comparing the meq of each element to the CEC. The meq of the cation element divided by the CEC, multiplied by 100 equals the percent of the CEC that the cation occupies. For example, if the CEC of a soil is 10 and the meq of potassium is 0.3, then $0.3 \div 10 = 0.03 \times 100 = 3\%$. The ratio of the percent potassium, magnesium, and calcium (K:Mg:Ca) represents the *base saturation*. Base saturation is the balance of base cations in the soil. Labs that offer base saturation values calculate these percentages automatically and usually offer ranges within which those values should fall. Typically, those ranges are 2–7% for potassium, 10–15% for magnesium, and 65–75 % for calcium.

Some managers assign more importance to base saturation than others do. There is strong evidence that excessive saturation of Ca and Mg can force K into the soil solution where it is susceptible to leaching. There is also anecdotal evidence that a Ca:Mg ratio below 7:1 or above 10:1 can lead to an eclectic list of problems. The Ca:Mg ratio, however, is not calculated from base saturation percentages. To determine the Ca:Mg ratio, divide the parts per million value for exchangeable Ca by the parts per million value for exchangeable Mg. For example, if analysis shows 1,450 ppm Ca and 183 ppm Mg, then $1{,}450 \div 183 = 7.9$ (to 1).

Base saturation values significantly above the suggested range may not be cause for concern unless they are creating an imbalance in the form of deficiencies. Only deficiencies should be addressed and by doing so, excessive saturation of other cation elements are often reduced. The CEC and base saturation tells two important details about the soil. First, it tells how much potash, magnesium, and calcium the soil can hold and second, it tells the balance of those nutrients.

Correcting cation deficiencies, in most cases, should be done in accordance with the CEC. In other words, one should not apply more cation nutrients than the soil can hold. Determining how much to apply can be easily calculated by multiplying the CEC by the desired saturation percentage and then multiplying that product by the cation's meq factor (see Table 4.2). For example, if an analysis reports the CEC of the soil is 6.3 meq/100 gr, exchangeable potassium is 24 ppm, and the base saturation of K is 1%, how much potash should be applied? If 4% potassium saturation is our goal then $6.3 \times 0.04 \times 390 \approx 98$ ppm K. In this equation, 6.3 is the CEC, 0.04 is equivalent to 4%, and 390 is the meq factor for potassium. Since 24 ppm already exists, then an application of $(98 - 24 =)$ 74 ppm K is appropriate. Conversion is also needed from ppm K to pounds per acre potash as fertilizers guarantee soluble potash, not elemental potassium. The conversion factor is 2.4 (see Table 4.3) so, $74 \text{ ppm} \times 2.4 \approx 177$ pounds per acre or ~4 pounds per 1,000 ft^2 of actual potash needs to be applied. The formulas for calculating Ca and Mg applications are:

$$\text{ppm Ca} = \text{CEC} \times \%BS_{Ca} \times 200 \quad (\%BS_{Ca} = \text{the desired base saturation of calcium})$$

$$\text{ppm Mg} = \text{CEC} \times \%BS_{Mg} \times 120 \quad (\%BS_{Mg} = \text{the desired base saturation of magnesium})$$

To convert parts per million Ca or Mg into pounds per acre, multiply by 2. Keep in mind, if the lab overestimates the CEC because of inappropriate extraction procedures, these formulas may not provide the most efficient results.

Cations are balanced by anions, which are ions that carry a negative electromagnetic charge. Cations, anions, and the exchange system in soil are crucial components in the cycles and chains of life on earth. Their importance is comparable to a vital organ which, by itself, does not sustain life, but without which life could not exist. Because an-

Table 4.3. Conversions

To Convert[a]	To[b]	Multiply By
ppm Potassium	ppa Potash	2.4
ppm Calcium	ppa Calcium	2
ppm Magnesium	ppa Magnesium	2
ppm Phosphorus	ppa Phosphate	4.6

[a]ppm = parts per million.
[b]ppa = pounds per acre.

ions have the same polarity as soil colloids, they cannot be held or exchanged by clay or humus particles. Important anionic compounds related to plant nutrition include nitrate (NO_3), phosphate (PO_4), and sulfate (SO_4). Phosphate ions can bond easily with several different soil elements or compounds, and do not migrate very far before developing a serious relationship with another soil chemical or they are assimilated by soil organisms or plants. Plants use phosphorus as a phosphate combined with hydrogen. Applications of highly available phosphate can, however, be eroded with normal surface runoff. This runoff can cause nutrient loss and possible eutrophication of waterways.

Phosphorus tests are inherently inaccurate because there are so many variables in the soil that affect the availability of phosphate. If deficiencies of P are found, however, then an application of some type of phosphatic material is usually warranted. The problem occurs when acceptable levels are found but soil conditions are such that it cannot become available to plants. The chemicals used to measure phosphorus in a soil sample attempt to mimic the natural chemicals produced in a typical soil. The variations inherent in soils from one square foot to the next, let alone regional or continental variations make it difficult to pinpoint typical conditions. Acids and other solubilizing substances created by soil organisms, plant roots, humic material, and soil chemical reactions are instrumental in freeing up phosphate by mineralizing organic phosphate from the residues of plants and other organisms and by biochemical reactions with soil compounds that contain phosphate. Ample soil organic matter is the key to a healthy and diverse population of organisms and adequate reserves of available phosphate.

Many labs report both available and reserve phosphorus and usually use symbols such as VL (very low), L (low), M (medium), H (high), and VH (very high) to indicate how the results of the analysis compare with what they consider to be acceptable levels. These symbols are also used to indicate reserves of other nutrients as well.

Nitrates and sulfates are susceptible to leaching. Aside from the issue of groundwater pollution, the loss of these nutrients, when applied as inorganic compounds, can be significant especially in well-drained, sandy soils that are typical of golf course greens. Soils with a greater capacity to retain soil solution can hold dissolved nitrates and sulfates more efficiently. This is not to say that a heavy soil is preferable. In heavy or compacted soils where the oxygen supply is limited, reduction (the biological or chemical removal of oxygen from compounds) can occur. Soil organisms (mostly bacteria) can reduce NO_3 or SO_4 to satisfy their own oxygen needs, which often transforms these nutrients into gases that can escape into the atmosphere. Sandy soils with enough organic matter to hold an ample reservoir of soil solution without compromising drainage is the ideal.

Tests for nitrates or sulfates are often optional, that is, not included on a lab's standard soil analysis. Nitrate analysis is a snapshot test because levels constantly fluctuate. Factors that influence nitrate content in the soil include soil temperature, carbon to nitrogen ratio, soil moisture, organic matter content, precipitation, biological activity, and the superintendent's fertility program. Samples from the same spot taken days apart can reveal significantly different levels of nitrate. If a nitrate deficiency (or superfluity) is suspected, soil samples should be drawn, dried, and sent to the lab via overnight delivery to get the most accurate results. Some managers refrigerate or freeze the sample and then ship it in an insulated pouch. The longer the time period between sample extraction and analysis, the more likely that changes in nitrate level will occur. Nitrate testing is more appropriate in arid and semiarid regions where the soil's source of water is primarily from irrigation and can be controlled. In the Northeast, where the weather can change rapidly, the results of a nitrate test may be meaningless. Tests for nitrate in sandy soils, especially where precipitation is unpredictable, are generally unreliable because of nitrate's propensity to leach. Acceptable ranges of nitrate vary in different types of soils but whoever performs the analysis should know. Leaf tissue analysis may be a better indicator.

Testing for nitrate as part of a N management program may be profitable on large areas where significant applications of fertilizer are proposed. The appropriate timing of the test, however, may create a practical problem. Samples for nitrate testing should be drawn after the soil has warmed up and the biological component of the soil has become active again. This may not occur until mid- to late spring in some areas. It may be inappropriate or inconvenient to postpone N fertilization this long. If N fertilization cannot wait, then applications of nitrogen can adulterate the samples that are drawn later for nitrate analysis.

Some labs offer a calculated nitrogen analysis called *Estimated Nitrogen Release* (ENR) based on the percentage of organic matter found in the sample. ENR (*expressed in pounds per acre*) is usually calculated using one of several conditional equations.

- Soils with a cation exchange capacity (CEC) of > 7 meq/100 gr and percent organic matter (% OM) < 0.4% then ENR = 20 × % OM + 24. If % OM > 0.4% then ENR = 20 × % OM + 2 6.
- If the CEC > 4 meq/100 gr but ≤ 7 and % OM < 0.4% then ENR = 20 × % OM + 39. If the CEC > 4 meq/100 gr but ≤ 7 meq/100 gr and % OM > 0.4% then ENR = 20 × % OM + 41.
- If the CEC < 4 and % OM < 0.4% then ENR = 20 × % OM + 49. If the CEC < 4 and % OM > 0.4% then ENR = 20 × % OM + 51.
- Reduce ENR by 20% in northern soils because cooler average temperatures inhibit the biological activity needed to release N from organic matter. ENR never exceeds 140 ppa (~3.2 lb per 1,000 ft^2).

It is difficult to predict how much nitrogen will actually become available from organic matter, however, because of the large number of variables that affect its release. These variables include soil temperature, soil moisture, soil atmosphere, and the form and stability of the organic matter found in the sample. Peat, for example, has very little N to offer plants or soil organisms. On the other hand, nitrogen applications should not be based solely on plant needs either. The N that turf receives from organic matter, especially when clippings are recycled, is significant. Clippings can supply up to one-third of the N that turf needs and the utilization is so efficient that less than 5% of its N is lost even in the worst of conditions. Organic matter is often a warehouse of nitrogen. Residing in every 1% of organic matter in 1,000 ft^2 of soil ~6 inches deep, is an average of ~23 pounds of nitrogen. A soil with 4% organic matter might contain over 90 pounds of nitrogen per 1,000 ft^2. This may sound excessive but normally, very little organic N is available at any given time. ENR estimates may sometimes seem higher than what a steward might want to rely on, but they should not be ignored by any means. If only one- to two-thirds of the ENR value is considered reliable and factored into a fertility program, the reduction of applied nitrogen is significant.

Sulfate is not as transitory as nitrate and few turf managers worry about sulfate deficiencies. Since the dawn of the industrial revolution, pollution from coal- and oil-burning facilities has deposited ample sulfur on the soil. As cleaner fuels have been introduced and more air pollution controls developed, less sulfur is being generated that might eventually enrich the soil. Given the drainage in sandy greens, the lack of organic sulfate reserves, and the propensity for sulfate to leach, occasional sulfate analysis may be in the best interest of the superintendent. The optimal range for sulfate is ~7–13 parts per million. If the lab reports results as ppm sulfur, multiply by three to convert sulfur to sulfate.

Most labs offer recommendations either automatically or optionally. This advice is based on interpretation of the soil sample's analysis. Too often, however, the lab technician who tested the sample doesn't know squat about the course compared with the person who manages it everyday. That manager should be able to interpret the analysis too. If the lab only gives recommendations based on ranking values such as very low (VL), low (L), medium (M), etc., but no actual data are available, there isn't enough information for the superintendent to interpret the test. The tremendous number of different analytical labs and methodologies makes it difficult to outline an interpretation method but there are basic tenets that can be used to interpret the data from most labs.

A lab that doesn't measure the pH of a client's soil sample would indeed be rare. The value of that analysis, however, can only tell us whether or not we need to lime (or acidify)—not how much to apply. There are various methodologies for determining lime recommendations that include a buffer pH test, a reactive aluminum test, titration of soil acidity, or buffer estimations based on soil organic matter, soil texture, or both. A buffer pH test usually gives a value of H in meq per 100 gr of soil and it takes ~½ ton (1,000 pounds) of pure calcium carbonate ($CaCO_3$) per acre to neutralize each meq of H. Legally, lime has to be labeled with a calcium carbonate equivalent (a.k.a. total neutralizing value). This percentage indicates the degree of purity compared with $CaCO_3$. If, for example, a liming product is labeled 95% calcium carbonate equivalent (CCE), then it will take 1,053 pounds to neutralize one meq of H per acre. The formula used to find this equivalent is, WEIGHT ÷ CCE = Equivalent (Note: percent = value ÷ 100, e.g., 95% = 0.95). The lab's target pH may be higher than the superintendent's. Knowing how to determine the degree of neutralization enables the superintendent to tweak the recommendations if necessary.

Some labs will perform a reactive aluminum test to determine lime requirements; however, this may be an inappropriate test for many golf courses, especially for turf that is established in very sandy soil. Lime recommendations based on reactive aluminum are often higher than those determined from a buffer pH test. It might be advisable to retest the pH before too many applications of lime are made.

The recommendations from labs that perform titration tests are difficult to check but may be a more precise method of calculating lime recommendations especially on sand greens. Sandy soil holds very few cations, including hydrogen, so buffer solutions are often not sensitive enough to give accurate readings. Titration is done with a water-soluble base, such as sodium hydroxide. The base is slowly added to the sample until the desired pH is achieved. Then the calcium carbonate equivalent to the soluble base can be calculated.

Buffer estimations based on soil organic matter, soil texture, or both are often as accurate as any other. The savvy superintendent knows a sand-base green requires a lot less lime than a clay loam in the rough to achieve the same change in pH. Lime applications of 10 pounds per 1,000 ft^2 or less may be adequate to achieve the desired pH. Light applications of lime while monitoring pH changes is a good way to familiarize oneself with the response characteristics of the greens. Time, however, should not be the criterion for liming. There is no reasonable way to predict the effect that time will have on soil pH. Too many factors are involved in pH changes and the only reliable method of determining the need for lime is with a pH test.

It is also important to use the appropriate type of lime. Calcite or calcium lime is mostly $CaCO_3$ but may contain some magnesium carbonate. Dolomite contains ~22% calcium and ~11% magnesium. The decision to use one or the other should depend on the existing Ca:Mg ratio. To calculate this ratio, divide the ppm calcium by the ppm magnesium shown on the analysis report. If the analysis gives values in pounds per acre, no conversion is necessary. If the result is less than seven, then using calcium lime is more advisable. If the result is more than 10, dolomite is probably a better choice. If the result is between 7 and 10, see if the lab offers ranking values such as very low (VL), low (L), etc. or if there is a bar chart that ranks the results. If Mg is ranked very low and Ca is ranked medium or higher then dolomite should probably be used. In other scenarios, calcium lime is preferable. Much of the calcium lime on the market (especially in the Northeast) contains up to 5% Mg so some Mg is still being applied.

It is relatively rare that an established golf course needs a heavy dose of lime (especially on greens), but if conditions occur that warrant an application of more than one ton per acre (~50 pounds per 1,000 ft^2), it is not advisable to apply all of the liming material at once. Spring and fall applications of not more than 50 pounds per 1,000 ft^2 can be made until the necessary amount has been applied. It is not advisable to apply liming materials during hot, stressful times of the season either.

Some soils may require acidification but it is important to understand why the soil's pH is elevated beyond optimum. If overzealous use of lime is the cause, it may be reasonable to slowly reduce the alkalinity with sulfur or some other acidifying material. If, however, the pH is elevated because of naturally calcareous soil or extremely alkaline irrigation water, pH reduction may be futile and attempts may cause more harm than good. Conditions such as these can be ameliorated to a certain extent by increasing the buffering ability of the soil with organic matter and the coincidence of biological activity. Turf can grow within a wide pH range but certain essential nutrients are often made unavailable when alkalinity (or acidity) reaches extremes. A more prudent approach to naturally alkaline soil may be to determine deficiencies (if any) through leaf tissue analysis and apply foliar applications of the missing nutrient(s).

Table 4.4 suggests application rates of elemental sulfur to lower the soil's pH. Rates should be reduced by one-third for sandy soil and increased by one-half for clay soils. Unless the application of sulfur is being thoroughly incorporated into the soil, it is not recommended that more than 20 pounds per 1,000 ft^2 are applied at one time. Sulfur bacteria use sulfur as energy and create sulfate ions (SO_4). Sulfate combines with soil hydrogen to create sulfuric acid, which reacts with carbonates in the soil to form carbon dioxide and water (and calcium sulfate). Removing carbonates in the soil lowers its pH. The soil temperature has to be above 60° F (15.5° C), however, for these reactions to occur because sulfur bacteria are relatively inactive below that temperature.

Cation exchange capacity is somewhat dependent on pH especially in soils where humus provides most of the colloidal exchange sites. This may be important when cal-

Table 4.4. Application Rates of Elemental Sulfur for pH Reduction[a]

	Desired pH				
Present pH	6.50	6.00	5.50	5.00	4.50
8.00	27	36	50	63	72
7.50	18	32	41	54	63
7.00	9	18	32	45	54
6.50		9	23	36	41
6.00			9	23	32

[a]Expressed in pounds per 1,000 ft^2.

culating potash, magnesium, or calcium requirements. If, for example, a lab reports ECEC instead of CEC and the soil is moderately acidic, it may be difficult to determine the amount of potash the soil can efficiently adsorb because as the pH rises so does the soil's capacity to hold potassium. As the pH rises, the difference between CEC and ECEC narrows. Once the soil's pH has been adjusted to the desired level, potash requirements can be calculated more easily based on CEC (or ECEC) and the desired saturation using the method described earlier in this chapter. This same method can be also used to determine if both calcium and magnesium are in balance. Given the lack of organic matter on a typical green, pH changes will not affect the CEC as much.

Many labs use ranking insignia such as VL for very low, L for low, M for medium (or O for optimum), H for high, and VH for very high. When ranking cation nutrients, most labs assign these values based on the soil's capacity to hold them. For example, if 70 parts per million of potassium were found in a soil sample that has a CEC of 3.0 meq/100 gr, the ranking might be H or VH even though there may not be enough to feed a healthy stand of turf. That same value (70 ppm), however, would be considered L or VL if the CEC of the soil were 12.0 meq/100 gr or higher. This may seem confusing but let's look at capacity in a different way. Envision a full can of soda next to an empty five-gallon pail. The capacity of the soda can is 12 fluid ounces and the pail holds 640 fluid ounces or ~53 times the capacity of the can. The level of material in the can is VH—it's full—but if one were to pour the contents of the can into the pail, the level in the pail would be VL. The ranking values for potassium, magnesium, and calcium are relative to the soil's capacity to hold cation nutrients. A sandy soil with a low CEC might show H or VH rankings for cations on the analysis report but these levels may still be inadequate.

Organic matter analysis can indicate the potential for biological activity and improved CEC but there are a lot of variations in organic matter that influence both the biomass and the cation exchange capacity. Undecomposed residues, for example, do not contribute to the CEC of a soil but can provide food for organisms, increasing their activity and the nutrients that they can make available to plants. Some residues, such as thatch, don't contribute much to either the CEC or biological activity. This material is more resistant to decay than residues such as clippings and because its contact with the soil is limited, organisms also have less access to it. Friable humus, such as compost, contributes a tremendous amount of resources for soil organisms but offers few exchange sites for soil cations. Stable humus, on the other hand, has much less to offer organisms but can adsorb a substantial amount of cations. The CEC contributions of organic matter can be measured to a certain extent. We know that the two soil particles

Table 4.5. Ca, Mg, and K Containing Materials

	Percent Analysis		
Material	Potash	Magnesium	Calcium
Sulfate of Potash (Potassium Sulfate)	52		
Potassium, Magnesium Sulfate	22	11	
Muriate of Potash (Potassium Chloride)	60		
Gypsum (Calcium Sulfate)			23
Epsom Salts (Magnesium Sulfate)		10	2
Calcitic Lime		3	33
Dolomitic Lime		11	22
Basic Slag		4	29
Kieserite		18	
Bone Meal		0.4	23

that adsorb most cations are clay and humus. If there is little to no clay in the soil, then humus is responsible for most of the soil's capacity to hold cations.

The cornucopia of other benefits organic matter can offer overshadow the importance of CEC derived from it. The energy stored in organic matter is what fuels the biological mechanisms in the soil ecosystem. One of the main objectives of this book is to convince the reader that the health and welfare of soil organisms is inextricably linked to the health of turf plants. With that in mind, the measurement of soil organic matter often indicates much more than its capacity to hold cations, which will be discussed later in this chapter.

The level of cation nutrients (i.e., potassium, magnesium, and calcium) found in a soil analysis may indicate a need for correction but the CEC indicates how much correction can take place at a time. Appropriate applications can be determined using the formulas mentioned earlier in this chapter. Once the amount of potash, magnesium, or calcium has been determined in pounds per acre (or per 1,000 ft^2), the application rate can be determined based on the percentage of nutrients contained in the fertilizing material. If, for example, one has determined that 4 pounds of potash are needed per 1,000 ft^2 and sulfate of potash (potassium sulfate) is being used. Sulfate of potash contains 51% potash (0–0–51), so $4 \div 0.51 \approx 7.8$ pounds of material per 1,000 ft^2. This same calculation can be used to determine calcium or magnesium applications. Table 4.5 shows the analysis of some materials containing potassium, magnesium, or calcium. The formulas (mentioned earlier in this chapter) used to determine cation nutrient needs based on desired base saturation indicate the amount of material that should be applied in accordance with the CEC. This, however, may not be enough for the needs of the plant, especially for turf growing in almost pure sand, which has a very low CEC. Increasing application rates beyond the soil's CEC may result in nutrient loss from leaching. Smaller, more frequent applications may be necessary. The advantages to a soil with a higher CEC are obvious but it requires significantly greater application rates to correct cation deficiencies. Imagine yourself filling the 12-ounce soda can and the 5-gallon pail mentioned earlier with water—using a hand pump. The pail has ~53 times greater capacity than the can. Not only does the pail hold a much greater volume, but it takes more energy to fill it. If both containers are filled to capacity, however, and the contents of both are used at the same rate, the advantages of the larger container are obvious.

Unlike potassium, magnesium, and calcium, phosphorus is not ranked in accordance with the soil's capacity to hold it. Phosphorus is measured by many different methods and the results are not always comparable. The acceptable range for one methodology may be completely different than for another. Most labs use the Bray method to determine both available and reserves of phosphorus in the soil. The weak Bray (P_1) extracts readily available phosphorus. The optimal range for most plants is 25–40 parts per million using the weak Bray test. The strong Bray test (P_2) extracts water-soluble phosphates, weak acid-soluble phosphates, and a small amount of insoluble phosphates. Adequate levels using this test range between 40 and 60 parts per million P. The Olsen Sodium Bicarbonate Extraction method is usually used on calcareous soil samples but is also appropriate for neutral to slightly acid soils that contain less than 3% organic matter. The acceptable range using this test varies depending on the pH of the sample. A range of 12–15 parts per million may be adequate at a neutral pH but 55–65 ppm would be needed at a pH of ~6.2. The Mehlich No. 1 (a.k.a. Double Acid Extraction) and the Morgan extraction methods are used primarily for low-capacity sandy soils with little organic matter and a pH of ~6.5. Optimum levels of P range from 20 to 50 parts per million extracted with these tests. There are still other acceptable methodologies each with their own optimum ranges so it is important to know which method the lab is using and what the analytical technicians consider adequate. Keep in mind that these ranges are more important during the establishment phase for turf. Once established, turf is normally very efficient at removing phosphorus from the soil and levels of P below the optimum ranges mentioned above may be adequate.

Phosphorus is usually measured in parts per million and, like other nutrients, must be converted to pounds per acre in order to calculate application rates. Fertilizer, however, is always labeled for its phosphate content so conversion from phosphorus to phosphate is also necessary. To convert ppm phosphorus to ppa phosphate, multiply by 4.6. This factor includes the conversion of parts per million to pounds per acre (i.e., 2) and phosphorus to phosphate (i.e., 2.2951).

$$2 \times 2.2951 \approx 4.6$$

If, for example, it is determined that 20-ppm phosphorus is required to correct a deficiency, then $20 \times 4.6 = 92$ pounds of phosphate should be applied per acre (~2 pounds per 1,000 ft^2). If the phosphatic fertilizer contains 46% phosphate, then $92 \div 0.46 = 200$ pounds per acre or ~4.6 pounds per 1,000 ft^2.

Phosphorus is relatively insusceptible to leaching but can percolate down beneath the root zone in sandy conditions especially if there is a very low level of organic matter. Generally speaking, if the soil contains ample amounts of organic matter, not only will phosphorus not leach but there is usually an adequate reserve of P complexed into it.

The importance of organic matter and the biological activity it supports is the thesis of this book. Unfortunately, the inherent problem with organic matter analysis is that it doesn't tell us about its stage of decay or anything about the biomass. A substantial amount of undecomposed residues in the sample may result in a high measurement of organic matter but doesn't necessarily represent a high level of biological activity.

An experienced manager can employ analytical observation to determine some things about the soil organic matter. The color, feel, and smell of a soil sample can give a relatively accurate indication of its characteristics. A dark, crumbly sample with the smell of freshly turned soil is a good indication of well-developed organic matter and an active biomass; however, it is impossible to judge the diversity and vitality of soil organisms. Respiration is a better indicator.

Most soil organisms generate carbon dioxide and the volume produced is a good way to measure their activity. Woods End Research (see Sources and Resources for contact information) markets an on-site soil respiration test called Solvita® that measures biological activity. A soil sample is placed in a jar (provided) along with a gel-coated paddle and the jar is sealed with a screw-on cap. The gel changes color in response to the amount of carbon dioxide being produced by soil biota. After 24 hours, the color of the gel is compared with a color key that identifies the level of biological activity. This test will not work properly if the sample is too dry, too wet, too hot, or too cold. The sample should ball up when squeezed but no water should be expelled and the test should be conducted at room temperature (68–77° F) and out of sunlight. If the soil needs to be moistened or dried, samples should not be drawn for a couple of days after treatment. The sample should not contain thatch, crowns, or any other undecomposed residues. Solvita measures the active biomass as opposed to the resources upon which it is living. If the percentage of organic matter found from chemical analysis is high but the production of carbon dioxide measured by Solvita is low, then the organic matter found is not providing resources to the biomass for some reason. Limiting factors can include moisture, temperature, or both but can also include pesticides, salt fertilizers, irrigation water, or the physical condition of the soil.

Soil Food Web, Inc. and BBC Labs (see Sources and Resources for contact information) perform biological soil analysis. The tests that each lab offers differ but all are designed to measure the quantity and diversity of soil organisms in a sample. The larger and more diverse the soil food web in the soil, the better the regulation of plant nutrients and the more protection from pathogens. Tests that determine total and active bacteria, total and active fungi, protozoa numbers, nematode numbers and diversity, and mycorrhizal fungi give an indication of the balance and interactivity of soil organisms. Some labs can also perform the same assay for leaf surfaces to determine if disease suppressive organisms are colonizing the phyllosphere. These tests tell both the level of biological activity and the balance of organisms involved.

Turfgrass does not have an especially high need for micronutrients but all plants need some. A deficiency of any nutrient, no matter how little may be required, can hold back plant development and functionality. Testing for micronutrients is not usually necessary, but if problems exist that are not explained by a standard test, it may be necessary to have it done. Most labs will keep their customer's soil samples for ~30 days after they have analyzed it so if micronutrient analysis becomes necessary within that time span, drawing new samples may not be necessary.

Most micronutrients are cations, but labs do not normally report results in relation to CEC. The levels of micronutrients needed in the soil are so small that virtually any CEC will hold what is necessary. Great care should be taken when correcting micronutrient deficiencies. There is a very fine line between too little and too much. The lab that does the testing usually makes recommendations but their recommendations are only as good as the sample taken. Rusty or corroded tools used to draw samples can easily adulterate a micronutrient analysis. Because of the high zinc content, galvanized tools or containers should be avoided regardless of what their condition may be.

Iron is a micronutrient turf needs and is rarely deficient in most soils but if the pH of a soil is too high or biological activity is too low, iron can become unavailable to plants. Excessive levels of phosphates, zinc, manganese, or even thatch can also lock out iron.

Many managers apply commercial preparations of iron to compensate for these conditions. Turf color will usually improve when given small amounts of soluble iron (ferrous iron) but if conditions in the soil are such that iron is immediately tied up, then much of the iron applied will quickly become unavailable to plants. If these soil conditions exist and it is impractical to correct them, foliar preparations of iron or other susceptible micronutrients may be more efficient. Some superintendents will use iron as the weather turns colder to counter chlorosis caused by unavailable nitrogen. As the soil temperature drops, so does the biological activity that regulates the availability of N. Iron is also used to treat chlorosis when applications of nitrogen can exacerbate other problems, including certain turf diseases. The introduction of iron does not have an immediately apparent effect on the balance of nutrients in the soil and does not seem to alter the nutrient balance in the plant. Some scientists believe the iron merely coats the leaves, as a paint would, and causes a surface reaction that manifests itself in a dark green color. Iron, like many other nutrients, is abundant in soil (not sand) but most of it is tightly bound in mineral compounds and unavailable to plant roots. Biological activity on and around the root system can dissolve some of these bonds and make some insoluble micronutrients available. In general, the more resources that are available to soil organisms, the more soil-bound nutrients can become available to plants.

Sometimes, macro- or micronutrient deficiencies (or superfluities) occur in plants but a soil analysis does not reveal the problem. Some edaphic anomaly may be causing a nutrient deficiency even though there's not a shortage of nutrients in the soil. Lackluster turfgrass performance in soil with ample nutrients may indicate a root pathogen, some type of stress, or a problem absorbing essential nutrients even if they are available.

Leaf tissue analysis can be performed to determine if the turf is lacking something important. Most of the plant's physiological activity takes place in the leaves and changes in nutrient absorption by the plant are reflected in the balance of minerals in the leaf tissue. Keep in mind, leaf tissue analysis won't tell you if carbon is the limiting nutrient.

To submit samples for leaf tissue analysis, 30–40 blades from fresh clippings should be gathered and spread out to air-dry overnight. Avoid contamination from soil or other debris that can adulterate the test results. The samples should be packaged in a paper bag or cardboard box with ventilation holes punched through the walls and placed inside an appropriate shipping container. Labs generally request information that is necessary to provide a useful analysis. Most labs provide forms that need to be filled out and should accompany the samples. Leaf tissue analysis should be done in conjunction with a soil analysis so that the condition of the plant and the soil during the same time period can be compared. Conditions in the soil—especially in sand greens—and in the plant can change relatively fast. Tissue analysis performed months before or after soil tests may not accurately reveal the problem or problems.

Leaf tissue and concurrent soil analysis can provide information needed both to solve and prevent problems. If, for example, a tissue analysis indicates superfluous nitrogen in the plant, this condition may eventually lead to other problems such as thatch, disease, or insect encroachment. Or, analysis may reveal that the level of phosphorus in the plant is below the acceptable range even though there is ample P in the soil. This scenario is relatively common in cold soils or where there is very low biological activity. The analysis can also point out deficiencies of seemingly obscure micronutrients such as manganese, copper, or zinc. Leaf tissue analysis is a useful tool that can be instrumental in solving mystifying problems.

Physical analysis is normally more common during the construction stage of a golf course. The ideal texture, consistency, and particle size, especially on greens, can, at least in the short term, mitigate some potential problems. The consistency of a designed soil, however, will inevitably change over time. Natural phenomena that, from the perspective of the plant, improve growing conditions eventually reduce porosity so that more soil solution is retained. These changes are usually to the chagrin of superintendents who are trying to preserve superfluous drainage.

Physical analysis not only measures the percentage of sand, silt, and clay but also determines the diameter of sand particles, and the sphericity and angularity of the particles. The sand particle diameters are usually divided into seven different size categories ranging from 0.05 mm to 2.0 mm. The USGA recommends ≤5% silt, ≤3% clay and the rest sand. Table 4.7 shows their suggested ranges for sand particles. Physical analysis also measures particle density, bulk density, infiltration rate, total porosity, aeration porosity, capillary porosity, and organic matter. Suggested values from the USGA are shown in Table 4.8.

Some managers will perform physical analysis on established greens or other areas of the course to monitor changes, measure the effects of physical treatment, or justify renovation. Labs that do physical analysis want to see an undisturbed profile of the soil usually about 12 inches deep. One lab we interviewed (N.W. Hummel & Co.) suggests sharpening one end of a two-inch diameter PVC pipe and driving it ~12 inches into the soil. When the pipe is extracted, the sample is contained inside. The ends of the pipe are then packed with newspaper and sealed with tape and the entire assembly (pipe and all) is sent to the lab.

Texture analysis is slightly different than physical analysis. Texture analysis determines the percentage of sand, silt, and clay in a given soil. Those findings are plugged into a texture analysis triangle (Figure 4.5) to determine soil classification. The triangle is used by extending lines from the appropriate starting points, parallel to the sides of the triangle, which are counterclockwise to the sides where the lines began. Like tying shoes, the triangle is easy to use but more difficult to explain. If, for example, a soil texture analysis discovered 40% sand, 40% silt, and 20% clay, the first line would begin on the 40 mark of the *percent sand* side of the triangle, drawn parallel to the *percent silt*

FIGURE 4.5. TEXTURE ANALYSIS CHART

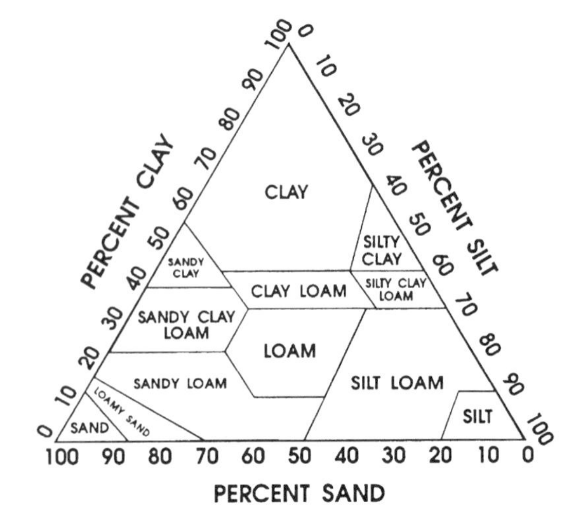

Table 4.6. Sand Particle Ranges

Screen Size	Percent Retained
2.0 mm (No. 10)	≤ 3% gravel
1.0 mm (No. 18)	≤ 10 combined
0.5 mm (No. 35) 0.25 mm (No. 60)	≥ 60%
0.15 mm (No. 100)	≤ 20%
0.10 mm (No. 140) 0.05 mm (No. 270)	≤ 5%

Table 4.7. USGA Physical Property Standards

Physical Property	USGA Values
Particle density	N/A
Bulk density	N/A
Infiltration rate	6–24 in./hr
Total porosity	35–55%
Aeration porosity	15–30%
Capillary porosity	15–25%
Organic matter	0.5–2.0%

side. The second line begins at the 40 mark on the *percent silt* side and is drawn parallel to the *percent clay* side. Where those two lines intersect is the texture classification of the soil. The third line may be drawn to complete the triangulation, but it is not necessary unless the first two lines intersect directly between two classifications.

A simple, on-site texture analysis can be conducted by collecting a soil sample in a clear graduated cylinder, test tube, or other graduated container and mixing it with an equal volume of water. Tap the container on a hard surface to compact the soil slightly and measure the volume of the soil sample before water is added. After adding water, seal the container and shake it vigorously for ~60 seconds. Then place it upright in a location that will allow it to be undisturbed for ~24 hours. The sand particles will settle to the bottom and the silt above that. Clay will eventually settle on top but the finer particles can take weeks to precipitate out of suspension. This is why we measure the dry sample before adding the water. If the volume of the original sample is known, it is easy to determine the percentage of sand and silt by measuring each layer and calculating what portion it is of the whole sample. To determine the percentage of sand or silt, divide the level of each by the total volume of soil and multiply by 100. If, for example, our soil sample measures 50 mL and we find that 20 mL of that is sand, then $20 \div 50 \times 100 = 40\%$. Once the percent sand and silt is determined, the texture analysis chart can be used to classify the sample. Physical and texture analyses are valuable tools. Familiarity with the soil one is working in allows for more informed decisions about irrigation, core aeration, liming, fertilization, and applications of pesticides.

Water analysis is another important test if the superintendent relies on irrigation. Insufficient water can quickly become the most limiting factor in turf performance but contaminated water can slowly create enormous problems that may not show symptoms until it is too late. Labs that test the irrigation suitability of water commonly measure sodium, calcium, magnesium, chloride, conductivity, sulfate, nitrate, pH, carbonate, bicarbonate, phosphorus, potassium, boron, dissolved solids, and sodium absorption ratio. These analyses can pinpoint most problems with irrigation water. There are, however, many other natural and man-made contaminants that can affect turf performance, and discovery can be difficult and expensive. Labs test for the presence of specific chemicals or conditions and can only find unknown contaminants by testing for different possibilities. If the list of those possibilities is long and if discovery occurs somewhere near the roster's end, considerable expense may be incurred. Water analysis is usually done when a new water source is introduced to the course but many managers will have their irrigation water tested periodically to monitor its quality.

It is important to keep analysis reports in perspective. The results of most tests are an average of the area being evaluated. The soil does not respond like a high performance engine where subtle adjustments can tweak out another two to three more horsepower. Fine-tuning the soil ecosystem is generally impractical. Not only are the results of most analyses inexact but factors that affect the physical, biological, and chemical conditions in the soil are constantly changing. If treatments are made with consideration for the entire soil ecosystem, then the whole solar-powered, biological turf-growing machine can benefit.

Chapter 5:

Pests

A pest is technically something that causes injury or damage to human concerns. On a golf course those concerns can range from mosquitoes strafing golfers to raccoons that get into the garbage behind the clubhouse. Pests that can cause severe damage to turf, however, are the greatest concern to superintendents. Severity is a subjective determination. It is the threshold of tolerance that the superintendent, greens committee, or golfers are willing to accept. At one time, damage that didn't interfere with play may have been considered acceptable but, over the years, aesthetic values have become an ever-increasing influence. On some courses, today's standards demand such intensive management that aside from photosynthesis, many of nature's contributions to the growing system are relinquished to management techniques. Artificial plant growth regulators do the job for natural hormone-producing organisms. Soluble salts of nitrogen and other immediately available nutrients do the work for free-living nitrogen fixers and organisms that process organic residues into plant nutrients. The job performed by predators is (sometimes) accomplished with insecticides that too often harm the predators. Fungal pathogens need only fear fungicides because the natural populations of parasites, predators, and other disease suppressive organisms have left for more lucrative jobs in other environments. And as these fungi develop their tolerance to fungicides, they can be even more successful. Earthworms, microarthropods, and other beneficial saprophytic organisms have moved to escape the grim realities of this chemical warfare. In reality, most of these impacted groups of organisms don't migrate to less hostile environments—they just don't survive. This micromanagement technique might not create so many problems if we fully understood the complexities of turf's ecosystem—but we don't. We do understand that the relationships that plants have with the biological component of their environment are important for their survival. If materials are applied that suppress the functions of these organisms, problems that require other chemical remedies often appear.

Turf—or should we say the superintendent—can become dependent on chemicals as easily as people can become dependent on drugs and like drugs, the more often that chemicals are used, the more often they seem to be needed. Although the golf course is largely a man-made system, it can be, nonetheless, a system that functions with balance and ecological grace. The occurrence of turf pests in large numbers might be a message to the superintendent that their ecosystem is somehow out of balance. Sometimes the message is simple: *You can't grow that here.* Other times the problems speak of infertility, excessive nutrient(s), compaction, poor drainage, excessive drainage, inadequate levels of organic matter, mechanical injury, or incorrect pH. Treatment of the symptom, that is to say, the weed, the insect, or the disease, does not necessarily recreate balance—

The greens at co-author Luff's course are cut at one-fourth inch and are admittedly slower than greens mowed at a lower HOC but stress related

it only eliminates the symptom, which nature will often replace with another. Additionally, applications of many pesticides often eliminate more than just the target organisms. The subsequent depletion of other organisms that may have functioned beneficially can create more unwanted symptoms. This cycle is a relatively classic scenario in turf and it can eventually result in a dramatic reduction of biological diversity and, consequently, grass quality in the immediate environment. Many of these changes may be unnecessary. The cause of the original problem might have been simple but it is complicated by multiple treatments that have further altered the soil ecosystem.

Grasses are naturally resilient and competitive organisms able to withstand the rigors of grazing animals. Grown under optimal conditions, grasses can compete with weeds and endure the occasional pathogen or small herd of grazing insects. Soil organisms also react to optimal conditions in a similar manner. Huge populations of soil dwellers compete for resources in a ecological orchestration that maintains a functional balance throughout the soil system. Infestations of pests have greater difficulty becoming established in an environment where so much competition exists. Conditions that harbor pest problems are often optimal for the pest, not the health of the plant. Low levels of organic matter, inappropriate height of cut, shallow root systems, excessive thatch, soil compaction, poor fertility, overirrigation, or incorrect pH can all contribute to pest problems by not providing adequate conditions for grasses to be healthy and competitive.

Investigations from the USDA Agricultural Research Service have shown compelling evidence that plants can cultivate populations of specific soil-dwelling organisms in the rhizosphere. They accomplish this by releasing special pabulum through their roots that attracts and nurtures these select groups. These organisms colonize the roots and can produce toxins that inhibit the growth of specific pathogens. They can also correct some nutrient deficiencies by making mineral ions, necessary for plant growth and health, available from the soil, and they can produce invaluable substances that regulate plant

problems are significantly less frequent. Herbicides and insecticides are never used and fungicide use is typically less than 1/10th of the regional average.

growth and help the plant develop resistance to environmental stress. These substances include amino acids, antibiotics, auxins, cytokinins, enzymes, giberillins, indoles, and vitamins. Some of these substances can be transported intact through the vascular system of the plant to help growth, resist stress, and defend against many pathogens and other foraging pests.

The side effects of many pesticides can include suppression, depletion, eradication, or an alteration in the balance and type of organisms in the rhizosphere. If symptoms of this biological malaise begin to appear, more pesticides are often applied and the problem worsens. Some biologists believe that, in the long-term, many pesticides can have subtle residual effects lasting for years and that some pesticide residues may combine to form new chemicals with unpredictable effects on soil organisms.

This isn't a claim that once a superintendent has begun using pesticides, there is no turning back. Biological health and balance can be restored but it is difficult to do this at the same time that the balance of soil organisms is being altered or destroyed with frequent applications of pesticides. Applications of concentrated, salt-based fertilizers can also impact developing populations of these select groups of root-dwelling organisms by affecting the osmotic pressure of surrounding soil moisture and the soil's chemical balance. Certain kinds of organic matter added to or inherently in the soil can mitigate many of the harmful effects on soil organisms from both pesticides and salt-based fertilizers providing the chemicals aren't being applied in overwhelming quantities (see Chapter 2, Fertility and Chapter 3, Compost).

Scientists know that healthy plants are not completely defenseless against pest organisms. Plant species that have no defenses, either their own or from symbiotic organisms, would have become extinct long ago. Most species of plants can synthesize chemical defense compounds that inhibit pest activity. The synthesis of these organic chemicals depends largely on the availability of the necessary elemental components (i.e., fertility) and the plants' overall health, strength, and vitality. In some cases the introduction of a simple primary nutrient such as potassium can help the host plant create stronger defenses to certain pests. On the other hand, excessive applications of some nutrients, such as nitrogen, can significantly reduce the plant's ability to defend itself. Plant resistance to insect pests and diseases is manifested in four basic forms.

- The first is the manufacture of toxins that affect the biology of the pest;
- The second is its ability to make itself less attractive either by smell, taste, or nutritional value, forcing the pest to go elsewhere for sustenance;
- The third is the development of tolerances to pest damage, i.e., the ability to thrive in spite of the adversity; and
- The fourth is a physical resistance created by thickening the outer layers of both stems and leaves.

Often, a plant may exhibit all four resistances.

All of the defensive reactions mentioned above depend on optimum fertility and effective stress management. Fertility must include adequate supplies of carbon dioxide from biological activity in the soil. Plants cannot process nitrogen into proteins and enzymes without carbon and soluble forms of nitrogen in plant tissue are what attract pathogens and insects.

It is also known that many plants can sense imminent pest problems and alarm other plants. When nearby plants are attacked, natural plant hormones are often released, signaling adjacent plants to produce defense compounds in anticipation of attack. The production of these hormones and the implementation of defense mechanisms depend on a healthy plant growing in a sound soil ecosystem.

Soil organisms also have the ability to produce allelopathic chemicals. Researchers believe that there are as many natural toxins in the environment as there are different organisms. These toxins are usually formed by organisms such as plants, fungi, or bacteria, but some may be mineral in nature. Antibiotics, for example, are made by soil fungi and are toxic to many bacteria. In some soils, fungal organisms can maintain prominence by suppressing bacterial growth. Bacteria, on the other hand, also produce toxins that can inhibit populations of fungi. Bacteria dominated soils can naturally inhibit the establishment of many pathogens. Toxins are usually produced as metabolites of biological activity or as compounds formed from the decay of various organic materials. Some species of organisms produce compounds that are toxic to themselves, thus providing their own control for overpopulation. Substances that are toxic to one organism may be nutritious to another. The best balance of these natural toxins in the soil is both directly and indirectly related to resources that provide food and habitat for a large diversity of soil organisms. It is true that many of nature's toxins have been isolated, synthesized, and concentrated by scientists into biocidal substances called pesticides; however, introducing these substances into the environment in unnatural concentrations can cause an ecological chain reaction with unpredictable consequences.

Plants can display a variety of reactions to grazing pest organisms. The amount of organic nutrients released into the soil through plant roots generally increases, creating a relative increase of soil organisms in the root zone. Biological activity can, in turn, increase the amount of carbon dioxide and mineral nutrients available to the plant. Most plants respond to aboveground damage by increasing root growth and the amount of nutrients stored in the root system. Apparently, reducing the amount of nutrients located in the shoots and leaves makes the plant tops less nourishing and less attractive to graz-

FIGURE 5.1. GROWTH VERSUS DEFENSE

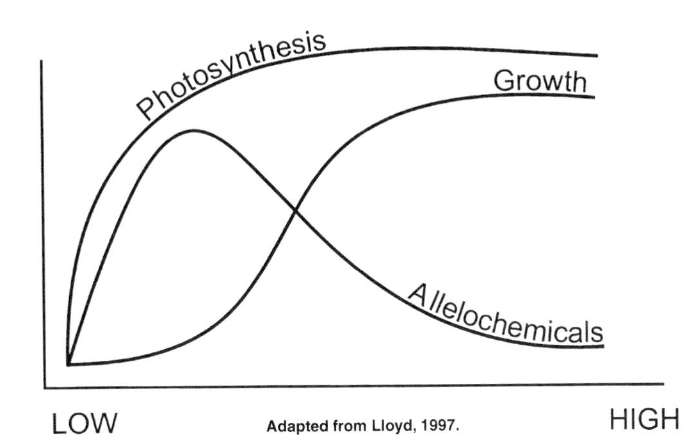

Adapted from Lloyd, 1997.

ing insects or pathogens. The stored nutrient in the root system increases the plant's ability to recover from pest damage. Plants can also exhibit a significant increase in the production of alkaloids, substances that are often toxic or distasteful to insects. When fertilizers are applied, however, especially concentrated, highly soluble salts that trigger lush growth, the production of these toxins is dramatically reduced. The more energy the plant has to expend growing, the less of it will be available to produce defense compounds (see Figure 5.1). A rich diet for turf can result in many different problems including disease, drought intolerance, and insect invasion.

Sometimes a superintendent's best efforts to produce a healthy turf are actually reducing turf quality by inflicting stress at inappropriate times. Mechanical injury caused by topdressing, core cultivation, mowing, rolling, or grooming is more difficult to overcome if heat, drought, or other environmental conditions are already stressing plants. Implementing these activities while ignoring the stressed condition of the turf is analogous to jogging with the flu. The same is often true for irrigation and fertilization. The more stress introduced to turf, the more likely pest problems will develop.

DISEASE

The simultaneous occurrence of a pathogen and a suitable host does not necessarily result in infection. Other conditions need to be favorable for pathogens to be successful. These conditions include but are not limited to temperature, humidity, precipitation (rain or snow), base fertility balance, pH, nitrogen management, irrigation management, height of cut, soil biological activity, and plant stress.

Turfgrass diseases that were once considered inconsequential have become major problems on the golf course due to excessive amounts of stress from traffic, low height of cut, sandy root zones, and other intensive management practices. After years of research, there is little doubt that stress is related both directly and indirectly to the vast majority of turf diseases. If the superintendent is attempting to manage disease with less or without fungicides, the mitigation of stress needs to be a high priority.

Like the human body, a plant has the ability to protect itself from diseases. Plants produce many different chemical compounds that are antagonistic to fungal pathogens. The ability of the plant to successfully resist disease depends on many different factors.

When stressed, the plant's normal metabolism is disturbed and compounds accumulate that cause the breakdown (oxidation) of lipids, proteins, and nucleic acids. Plants normally produce antioxidants to defend against this loss but production of antioxidants during stressful times is not always adequate to prevent damage. The plant responds by using lots of energy to create more lipids, proteins, and nucleic acids. However, photosynthesis occurring in the blades is often producing less energy than the amount being consumed because of stress factors like traffic, heat, and drought, or often the height of cut is the problem. If turf is mowed too closely, less leaf surface area is exposed to synthesize energy from the sun. As stress increases and energy production decreases, the plant becomes more defenseless against pathogens.

A substantial amount of research establishes a correlation between inadequate production of antioxidants and the occurrence of disease in turf. If the turf plant can produce adequate amounts of antioxidants during periods of stress, photosynthesis and metabolism will often be maintained at normal levels and the plant will be stronger, less susceptible to, and better able to defend itself against pathogen infection.

Researchers discovered that certain natural plant hormones, such as auxins and cytokinins significantly increase concentrations of antioxidants (such as alpha tocopherol, superoxide dismutase, and beta-carotene) in plants during periods of stress. These hormones are available in synthetic preparations but are also abundant in natural seaweed (*Ascophylum nodosum*) extracts. These extracts are commercially available in both liquid and dry-soluble preparations and can be used often and in varying concentrations without fear of overdosing. Most seaweed formulations are compatible with fertilizers or pesticides and can mitigate the ecological impact caused by some of these materials. In fact, researchers found that seaweed extracts can counter the osmotic effects of salt-based fertilizers. Seaweed can be applied as often as once per week and some studies suggest that when applied with fungicides, insecticides, or herbicides, the seaweed improves the performance of the pesticide. Apparently, this phenomenon is due to the seaweed's ability to improve plant health despite the pesticide's ecological impact. Pesticides are not always as discriminating as manufacturers would like us to believe and there is usually some ecological side effect from applying them. Researchers believe that seaweed minimizes the plant stress caused either directly or indirectly by pesticides.

Humic acids and humates have also been investigated and found to increase the plant's production of antioxidants. Research using combinations of seaweed and humic materials shows some very promising results.

Plants have strong defenses against disease but for these systems to function properly the plants must be healthy. The defense systems of over- or underfed plants, or plants experiencing severe stress are often not able to operate adequately. A strong plant allows only selective movement of materials through its tissue. The cellular structure of the plant is normally impermeable to anything plants want excluded. Under certain conditions, however, movement through and between cells changes from selective to passive. Cell walls are weaker and the tightly-knit arrangement of cells is loosened. This is often a result of too much nitrogen, not enough potassium, or both. It can also be caused by a long list of other conditions. Nitrogen can exist in plant tissue in soluble forms such as nitrate, ammonium, or amino acids. These forms are most attractive to pathogens and the loose-knit cellular structure makes it easier for penetration and infection. Once the plant has transformed soluble nitrogen into protein, it is much more difficult for pathogens to become established. It is harder for pathogens to assimilate and digest protein compared with more soluble forms of nitrogen. Plants need carbon, hydrogen, and oxygen to make protein from nitrogen. Their only source of carbon is from carbon dioxide in the atmosphere. In a rich soil, carbon dioxide is constantly being generated by soil organisms and plants have their best opportunity to absorb it as it seeps from the soil's surface. If inadequate amounts of carbon dioxide are evolved in the soil and too much nitrogen is applied, plants often become weaker and nutritionally more attractive to pests.

Carbon is also needed for the production of many defense compounds. Plants can produce enzymes like chitinase, which breaks down the penetrating component of pathogens. Carbon is also a component of lignin and callous materials that plants can produce to inhibit pathogen penetration.

Healthy plants can also produce bursts of hydrogen peroxide—a particularly lethal substance to most pathogens. Research has found that seaweed extracts, humic acids, or both can help plants increase production of many defense compounds.

Other protection from disease pathogens comes from the soil or more accurately, from soil organisms. As cliche as it may sound, there is truth to the axiom that a healthy soil promotes healthy plants. In the soil there are thousands of different organisms that exist for two common purposes—to survive and to reproduce. The most limiting factors in their existence are resources such as food, water, and a proper atmosphere. If these organisms were given an unlimited supply of resources, their populations would increase beyond the earth's spatial capacity in a matter of weeks. Organisms must compete with each other for the limited amount of resources in the soil. Most soil organisms can be categorized into one of three groups—parasite, saprophyte, or predator. The parasite is an organism that derives its energy from another life form, whereas a saprophyte derives its energy from the dead tissue or residues of other organisms. Many soil organisms—in fact, most turfgrass pathogens—can be both parasitic and saprophytic. As resources dwindle in a given environment, certain segments of the biomass may become dormant or die off, thus giving some surviving organisms a distinct advantage in terms of competition. In an environment depleted of organic residues, parasitic organisms tend to dominate. In fact, some saprophytic organisms may become parasites. Several important turfgrass pathogens exist as saprophytes until conditions prevail that allow this change to occur. If the dominant organisms happen to be plant pathogens, then a microbial imbalance may be manifested as plant diseases. Poor soil and other conditions can cause an organism to change from being saprophytic to pathogenic. Fusarium, for example, is most often a saprophytic fungi that decays organic residues but it can become a pathogen when it encounters poorly nourished roots. Unfortunately, its reputation is only that of a pest and it is usually treated with a fungicide, which can deplete or destroy other groups of beneficial organisms. Saprophytic organisms are valuable for their production of carbon dioxide, one of the most required nutrients in a plant's diet.

A rich soil, as its name implies, contains a wealth of resources. Many of us imagine a rich soil to be one that contains an abundance of plant nutrients capable of producing excellent crop yields. The term, however, has a deeper meaning. Resources are in many forms and benefit many different aspects of the soil system. Most of them are linked, either directly or indirectly, to one another and to an important soil component—organic matter. Although the percentage of organic matter in a mineral soil is relatively small, the function of this component is responsible for a lion's share of the system's benefits. The proteins, carbohydrates, humic substances, and other compounds that comprise the soil's organic fraction are resources for billions of soil organisms, many of which are disease suppressive.

Many superintendents have discovered that the incidence of disease diminishes significantly when the area is biologically enriched with well-aged compost (see Chapter 3, Compost). Apparently, the amount and type of organisms living in most mature composts are very antagonistic to pathogenic fungi. Some believe that many pathogens revert to a saprophytic existence when an area is enriched with compost but the reasons may have to do with plant health and not the availability of organic residues. Researchers from Ohio State University, in fact, discovered what they term "Compost-Induced Systemic Acquired Resistance" which describes a phenomenon that occurs in plants growing in a compost-based potting mix. Their experiment was conducted in containers where half of the plant's roots were separated from the other half by an impermeable wall through the center of a container. The peat-based potting mix on one side of the container was infected with *Pythium ultimum*, *P. aphanidermatum*, and *Colletotrichum orbiculare*. The other side of the container was filled with either a soil mix, a peat mix, or a compost mix. Apparently, the plant roots growing in the compost not only showed resistance to infection, but so did the rest of the plant. In the other containers, disease symptoms were evident on the roots in both sides of the container and on the aboveground portion of the plants. Many turf managers have found that soil management practices that encourage the development of organic matter, soil organisms, or both also have a similar effect.

The incidence of organisms is often related to the soil's organic matter content. Organic matter levels are controlled primarily by plant growth and climate. Cultivation practices such as tillage, aeration, irrigation, and liming, however, can accelerate the decomposition process (see Chapter 6, Cultural Practices). Plants are the main source of raw materials for the formation of organic matter, and the main influence for the development of microbial biomass. The roots of plants release up to 50% of the organic compounds produced by photosynthesis into the soil, creating an area of intense biological activity around the roots. This activity serves many different functions that affect the soil, plant, and atmosphere. Antagonism and competition directed at pest organisms are only a couple of them.

Cornell University has done a significant amount of research on the suppression of turf diseases with applications of aged compost. They have attempted to isolate the bacterium that is causing the disease suppression, but found that many species of microorganisms are responsible. In fact, they discovered that between 65 and 100% of microbes found in compost may have disease-suppressive characteristics. This discovery suggests that healthy soils, with ample organic matter, are also inherently rich with these disease suppressive organisms. Cornell's research was conducted primarily on golf course greens, where they would apply as little as 10 pounds of topdress mix, containing as little as 20% compost per thousand square feet five to seven times per year. They warn, though, that not all composts are created equal. The more effective composts for disease suppression are those that are aged for two years or more. They have also found that yard waste compost is relatively inferior for disease suppression. This is probably because the abundance of cellulose and lignin in yard waste does not support the types of bacteria needed for disease suppression. Ten to fifteen years of disease suppression studies using compost show, without a doubt, the potential of well-aged compost, made from the proper ingredients, to significantly reduce the severity of many different turf diseases that affect both the roots and shoots. Studies also show that the incorporation of high-quality compost into the green (during construction or renovation) can provide disease suppression for many years. Research also shows disease suppression from the use of compost teas, but the data are incomplete. Dr. Eric Nelson and other researchers from Cornell University formulated compost tea by mixing one part mature compost with five parts water and allowing it to soak for three to seven days. The liquid was then filtered and sprayed onto plants infected with pythium root rot. The results in the lab were far more encouraging than those in the field, and funding was exhausted before further study could be accomplished. Dr. Nelson felt that better results could have been observed if they were able to continue with the research, but that using high quality compost solids instead of the tea would always net better disease control. Dr. Elaine Ingham, an Associate Research Professor from Oregon State University, had better results using compost tea but hers was a much more complex production system which constantly incorporates air into the water so that valuable aerobic organisms can thrive. Her research (and others') demonstrated that compost tea, made correctly from well-aged compost (that is also made correctly), can be an important tool for turf managers against plant diseases. Brewers made specifically for the production of aerobic compost tea are available (Growing Solutions, Inc., Northwest Irrigation). Testimonials from superintendents who have used compost tea from an aerobic brewer have been very positive. These superintendents report the reduction and, more often, the elimination of fungicide use.

Inoculation with disease suppressive organisms is becoming a more popular practice with superintendents but applications often need to be repeated—frequently—because average soil conditions do not support large enough populations of these microbes to sustain disease suppression for a prolonged period of time. Additionally, many of the suppressive organisms used are UV sensitive and must be applied in the late evening or at night. There are a wide variety of organisms that have been found to be disease suppressive and although many have been tested—with positive results—few are both Environmental Protection Agency (EPA) registered and commercially available. Inoculants can be applied in dry material, sprays, mixed with topdress materials, and through irrigation systems. In fact, brewers that ferment large populations of disease suppressive strains are commercially available (Quarles, 1996). The brewers are connected directly to the course's irrigation system and usually deliver the microbes in the late evening to avoid exposure to ultraviolet light. One of the problems associated with using microbe inoculants is making sure an adequate population is not only delivered to, but also sustained in the turf. Commercially available organisms, such as *Trichoderma harzianum*, applied at labeled rates, should provide an adequate population as will many of the brewers, but distributing them evenly and sustaining adequate populations—a million or more microbes per gram of soil—is significantly more difficult. Soil environments with higher levels of humus and organic residues can sustain greater populations of soil organisms for a longer period of time but the conditions necessary to sustain the specific microbes in inoculants are unknown. The fierce competition for resources in the soil favors indigenous organisms but the diversity of those native groups also changes as environmental conditions vary.

Factors that influence the different groups of disease-suppressive soil and leaf-surface organisms include moisture, temperature, soil conductivity, pH, soil atmosphere, turfgrass variety, and organic matter. Introduced inputs such as fertilizers, pesticides, and topdress material also influence the soil's biota. Research at Oregon State University has shown that there are different biological environments in the soil that are very influential. A fungi dominated soil, for example, favors a forest environment, and a soil with a slightly higher population of bacteria than fungi favors grasses. It is not unreasonable to assume that conditions favoring a balance of bacteria and fungi can also favor disease suppression.

Inoculants found to be successful at reducing the severity of turfgrass diseases usually contain only one or two species of bacteria or fungi. The list of those organisms tested includes *Acremonium* sp., *Azospirillum brasiliense, Bacillus subtilis, Enterobacter cloacae, Flavobacterium balustinum, Fusarium heterosporum, Gacumannomyces spp., Gliocladium virens, Lactisaria* spp., *Phialophora radicicola, Pseudomonas aureofaciens, P. fluorescens, P. lindbergii, P. putida, Rhizoctonia* spp., *Serratia marcescens, Streptomyces* spp., *Trichoderma hamatum, T. harzianum, T. polysporum, Typhula phacorrhiza*, and *Xanthamonas maltophilia*. Of those listed, only a few are commercially available as EPA registered products but the list grows. Many others are commercially available (with new ones appearing on the market regularly) and, although most are effective, some have not been tested thoroughly, so it is advisable to contact the nearest land grant university that conducts turfgrass research for advice. The EPA has a web site that lists all of the registered biopesticides (http://www.epa.gov/pesticides/biopesticides/).

There is exciting new research from the University of Massachusetts about a new antifungal bacterium (accidentally discovered) that has been labeled APM-1. This organism produces powerful antifungal toxins that seem to affect most fungal pathogens and can be applied at much greater intervals than the brewed organisms mentioned above. The bacteria can exist in spore form and be delivered on a dry carrier to the turf canopy. There are usually not enough nutrients in the soil to sustain the growth of this organism but trials using an organic fertilizer as a carrier are showing positive results. Since spores can remain dormant for prolonged periods of time, a bagged product can have a shelf life of several years. Preliminary findings show sustained disease control for longer periods of time after a single application and that plant growth is significantly enhanced. This additional benefit is puzzling to researchers but they think the bacterium lives endophytically (within the tissue of the plant) and is somehow able to both protect and promote from the inside. The university, along with the New England Turfgrass Foundation, has applied for a patent on this new bacterium and availability may be in the not-too-distant future (Torello et al., 1999).

Biological control of diseases with inoculation is a viable alternative to fungicides. Handling and applying microbes often requires more attention and vigilance than fungicides but doesn't have the ecological impact of most fungicides. However, since there are so many different, native organisms that are able to suppress turf disease to varying degrees, it is logical that developing a resource-rich habitat for them in the soil would serve the superintendent well.

Some of our cultural practices may be responsible for altering the system's capacity to defend against diseases. Mowing is a stressful experience for grass, but mowing with a dull blade significantly increases stress. Turf mowed with a dull blade is often torn rather than cut, and the jagged wound left behind is an ideal place for pathogens to become established. Additionally, the extra stress inflicted upon the plant can lower its natural defenses against infection. The front face of the bed knife should be sharpened and the reels back-lapped every one to two weeks or more often if topdressing material or fertilizer causes dullness. If a bed knife or reel is nicked, it should be repaired immediately. Cleaning the reels or the underside of the mowing units is also an important practice. Removing potential inoculum from the mower can check the spread of pathogens and even some insect problems. Problematic greens should always be mowed last. Other stress-producing practices such as improper mowing, excessive traffic of maintenance vehicles, excessive use of pesticides, clipping removal, excessive applications of lime, poor drainage, and improper fertilization (especially with nitrogen) can significantly decrease the turf plant's resistance to many diseases.

Mowing too short reduces photosynthesis that occurs in the plant blades. The synthesis of organic compounds is essential for plant systems to function. Up to 50% of these compounds are exuded through the roots into the soil, nourishing large and diverse populations of organisms. These organisms are capable of, among many other things, disease suppression and the generation of carbon dioxide, a vitally important nutrient for plants. Mowing at a greater height is always beneficial for grass health, but player pressure at a golf course can keep the grass short—too short. Generally speaking, the lower the HOC, the higher the risk for disease. Many superintendents would consider greens mowed at one-fourth inch as shag, but they are much less susceptible to disease. Rolling the green can maintain better ball speed without lowering the height of cut. The leaf surface area of most greens-type grasses increases by 30% for every eighth of an inch the mower is raised. That means that 30% more photosynthesis can occur.

Excessive foot or vehicle traffic can cause soil compaction, which inhibits the movement of water and the exchange of gases between the atmosphere and the soil. This condition reduces the amount of oxygen available to roots and the amount of carbon dioxide available to blades, both of which cause stress to turf plants, decreasing their natural resistance to pathogens. Pythium root rot, summer patch, anthracnose, and other shoot and root pathogens gain an opportunistic advantage with compaction and poor drainage. Once disease organisms become established, traffic serves to further spread the problem to other areas. Areas subjected to excessive traffic need to be conditioned. The compaction needs to be relieved and a program for building soil organic matter should be considered. Topdressing with well-aged, good quality compost after core aeration is an excellent practice whenever and wherever possible (see Chapter 6, Cultural Practices). Research from Cornell University suggests that core aeration, by itself, offers only temporary relief of compaction problems. Aid from earthworms and soil-aggregating organisms is still needed, and resources such as compost are necessary for their populations to grow. Courses that see more rounds need larger greens (and tees) so that the wear and tear can be diluted.

Aeration is an extremely important practice. It reduces compaction caused by foot and vehicle traffic and immediately resuscitates the exchange of gases between the below- and aboveground atmospheres. Oxygen from above can reach the roots and the turf canopy can absorb carbon dioxide from below. On severely compacted greens, aeration may be analogous to finally getting a breath of air after swimming the entire length of a large pool—underwater. Proper aeration is a necessary cultural practice to control many turfgrass diseases but timing and frequency are critical. Aeration should always be done when the grass is thriving because, as helpful as it is, aeration still inflicts stress to turf. If the turf is already stressed by heat or drought and signs of disease are beginning to appear, aeration can exacerbate the problem even further. Once compaction becomes the cause of severe stress, the logical remedy—aeration—may be too late.

Another counterproductive side effect of aeration, especially core aeration, is the inadvertent depletion of soil organic matter (humus). Oxygen introduced into the soil increases root growth but also increases the oxidation of organic matter. Removal of cores may be a direct evisceration of the some of the best soil on the course. Cores that are collected can be saved and added to the compost pile. Core holes can be filled with a com-

FIGURE 5.2. PESTICIDE EFFECT ON DISEASE SEVERITY
(Brown Patch)

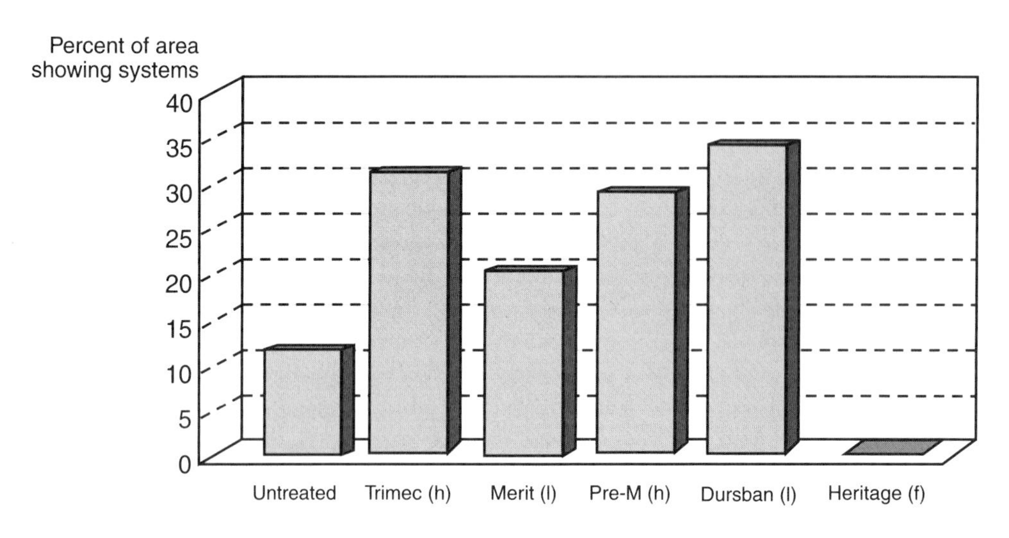

Adapted from Nelson, 1995.

post/sand topdress mixture to replenish the loss of organic matter and inoculate the green with disease suppressive organisms (see Chapter 3, Compost).

The excessive use of pesticides is another factor that can contribute to the severity of turf diseases. The purpose of pesticides is to kill biological entities that are injuring the health or appearance of cultivated plants. Unfortunately, many pesticides do not discriminate sufficiently between pest and beneficial organisms. They can alter the biological system enough to stress the preferred species of plants. Whether the product is a fungicide, herbicide, insecticide, nematicide, or rodenticide, there is evidence that other, nonpest organisms contributing to the well-being of the soil and cultivated plants are affected. The degree to which these biocides are used can impact some of the beneficial functions in the soil or plants, causing stress and presenting opportunities for disease organisms to become established. Researchers at Clemson University and Cornell found that a variety of different products designed to control weeds and insects increased the severity of brown patch (see Figure 5.2). It is not unreasonable to assume that the pesticide is decreasing the plant's ability to resist infection. It is also likely that stress inflicted by many pesticides might give an advantage to many other pathogens. In fact, some field studies show that diseased plots treated with fungicides had more severe symptoms at the end of the spray interval compared with plots that were not treated at all.

The vast majority of fungal organisms in the soil function beneficially. They perform such tasks as the decomposition of organic residues and the generation of carbon dioxide—a vital nutrient for photosynthesis. Some, such as mycorrhizae, have symbiotic relationships with plants that offer access to water and nutrients otherwise out of reach from the plant's roots. Others are parasitic or antagonistic to many plant pathogens. Mycorrhiza is a family of beneficial fungi that colonizes plant roots and not only provides nutrients and water for plants but also protects them from root pathogens like pythium. Most fungicides, however, do not discriminate between these organisms and can destroy more beneficial fungi than pathogens. It is not a logical strategy. The use of nematicides to control phytopathogenic nematodes will also kill beneficial nematodes, many of which are vital to plant health or antagonistic or parasitic to fungal pests. The Labronema nematode, for example, lives on many different fungal organisms, some of which may be pathogens.

A secondary but significant price paid for using certain fungicides is earthworm mortality. The suppression or decimation of earthworm populations from fungicides can eventually cause other problems, such as thatch buildup, soil compaction, reduced water infiltration, and the reduction of plant vigor. Earthworms are sometimes considered a nuisance by superintendents because of the casting piles left on the surface of the green. The benefits they provide, however, far outweigh the inconvenience of the castings, which can be easily dispatched (see Chapter 6, Cultural Practices).

Plant pathologists recommend removing the clippings from diseased areas so pathogens are not spread into other areas, but research suggests that leaving healthy clippings behind can suppress many disease organisms. Apparently, some of the organisms that decompose clippings are antagonistic to many plant pathogens. Additionally, the recycling of nutrients and organic matter back to the soil promotes the growth of healthier plants that are better equipped to defend themselves against diseases. Pathogens can be spread by foot and vehicle traffic, mowers, wind, irrigation, and even by birds and insects. The key defense is a soil rich in microbial life, capable of suppressing pathogen infection. If greens are mowed early in the morning when the grass is normally wet from dew, clippings usually have to be collected or they form clumps. If late afternoon mowing can be arranged when the grass is dry, clippings can often be left behind without causing problems with players. Coauthor Richard Luff reported a discernible decrease in the incidence and severity of disease when greens were mowed in the late afternoon and clippings were left behind. Note: Mowing severely heat-stressed plants is never a good idea. If the clippings must be collected, their final destination should be the compost pile so that, eventually, the compost can be returned to the green, tee, or fairway for its disease suppressive characteristics (see Chapter 3, Compost).

Proper pH and nutrient balance can also play a major role in the management of diseases. Stress associated with nutrient deficiencies or incorrect pH can make the difference between a turf that resists infection and one that does not. Testing the soil, leaf tissue, or both on an annual basis is an excellent way to monitor the balance of nutrients in the soil and plant (see Chapter 4, Analysis). Proper fertility is essential for a healthy, disease resistant, stand of turf. Disease resistance depends on an adequate supply of all essential nutrients. Excessive applications of nutrients or lime, however, can be as detrimental to plants as deficiencies are. Time should never be the criterion for applying lime, only a pH test. Different soils have different capabilities to hold nutrients, and the optimal capacity of one soil may be twice that of another. These holding capacities are usually measured and considered by soil testing laboratories making fertility recommendations. Exceeding those recommendations can cause as many (or more) problems as not applying enough. Attention to the calcium:magnesium ratio in the soil analysis may also increase disease resistance. Many believe that ratios lower than 7:1 can make soils more prone to compaction and, subsequently, to disease as well. If a soil analysis indicates a low calcium:magnesium ratio but lime is recommended, dolomitic lime should be avoided (see Chapter 4, Analysis).

Some turf diseases are sensitive to a high or a low soil pH and some scientists suggest raising or lowering the pH on greens that are prone to infection by these acid or alkaline intolerant pathogens. Unfortunately, other organisms—including turf plants—may also be sensitive to changes in pH. Many of those organisms may be beneficial, perhaps even disease suppressive. A lower pH generally inhibits bacteria but has little effect on fungal organisms. Bacteria typically antagonize pathogens more than any other group of soil organisms. Optimum pH for turf and soil organisms is logically best for disease suppression. Healthy plants and a balanced biomass will yield the strongest defenses against disease. In a naturally alkaline soil, it may not be practical to lower the pH. The amount of sulfur or other acidifying material needed to reduce the pH to an optimum level can cause more problems than it solves. Greens where calcareous sand has been applied year after year may have pH anomalies that cannot be completely corrected (see Chapter 4, Analysis).

Diseases that are more successful where fertility or pH is either inadequate or excessive include Rhizoctonia diseases, Pythium diseases, bermudagrass decline, Leptospaeria spring dead spot, Curvularia, dollar spot, zoysia rust, Bipolaris leaf blotch, yellow patch (winter patch), brown patch, Fusarium, take-all patch, summer patch, necrotic ring spot, snow molds, snow scald, red thread, pink patch, anthracnose, brown blight, melting out, powdery mildew, rust diseases, stripe smut, downy mildew, gray leaf spot, fairy ring, and virus diseases.

Proper nitrogen management is especially important. Too much or too little nitrogen encourages the vast majority of the above-mentioned diseases. The plant's use of nitrogen varies because of many different factors including soil texture, organic matter (quantity and quality), precipitation (or irrigation), soil and air temperature, turf variety, root mass, clipping management, cultural practices (height of cut, verticutting, aeration, topdressing), solar exposure, and the type of nitrogen applied. Even if best-guess nitrogen recommendations were correct, they don't take into account any of the aforementioned conditions. Considering the different factors that influence turf's assimilation of nitrogen is important if the superintendent wants to control diseases with less or without fungicides. It is also extremely difficult. Biological management of nitrogen can be a much more efficient method of regulating its availability to turf plants (see Chapter 2, Fertility).

Dr. Gail Schumann from the University of Massachusetts produced a chart (see Table 5.1) outlining cultural controls of specific turf diseases. Seventy-five percent of the recommendations involve either increasing or decreasing soil fertility. The chart suggests that too much or too little soil fertility can encourage plant pathogens. Common sense dictates that optimum soil fertility cannot be improved upon, but achieving this ideal is nearly impossible. There are too many variables that affect the availability and the plant's need for nutrients. At an ecologically maintained golf course, most nutrients are applied in a form that is unavailable to plant roots. The release of these nutrients is facilitated by biological activity in the soil, regulated by many different environmental conditions such as climate and soil moisture. Coincidentally, the conditions that encourage this activity also encourage the growth of plants. This natural coincidence synchronizes the availability of nutrients from natural organic and natural inorganic sources with plant need, and very often regulates the release of those nutrients for optimum plant utilization. In addition, the biological activity stimulated by organic resources can compete with, antagonize, or parasitize many turf pathogens. Where she suggests lower-

Table 5.1. Cultural Controls for Certain Turfgrass Diseases[a]

Anthracnose	Fertilize, aerate, raise HOC, reduce water on leaf blades.
Brown patch (rhizoctonia blight)	Avoid excess N and water, minimize leaf wetness time.
Dollar spot	Fertilize, aerate, minimize leaf wetness time, use resistant cultivars.
Fairy ring	Core and water, mask symptoms with N or iron, in severe cases—remove soil.
Fusarium leaf blight, crown and root rot	Avoid drought, minimize leaf wetness time, reduce thatch.
Leaf spot melting out	Avoid excess N and water, minimize leaf wetness time, raise HOC, use resistant cultivars.
Necrotic ring spot	Avoid water and fertility stress, aerate, reduce thatch, use resistant cultivars.
Powdery mildew	Improve air flow, reduce shade, avoid excess nitrogen.
Pythium blight	Avoid excess N, improve drainage, don't irrigate at night or mow in wet weather.
Pythium root rot	Improve drainage, aerate, raise HOC.
Red thread/pink patch	Fertilize, avoid low pH, minimize leaf wetness time, use resistant cultivars.
Rust	Fertilize, aerate, avoid water stress, minimize leaf wetness time, use resistant cultivars.
Slime molds	Minimize leaf wetness time, hose or rake away mold.
Snow molds	
Typhula blight (gray snow mold)	Let turf go dormant, mow until growth stops, minimize snow cover time.
Fusarium patch	
Stripe smut	Purchase smut-free seed, avoid excess N in spring, avoid water stress in summer, use resistant cultivars.
Summer patch	Raise HOC, lower pH if possible (see also Necrotic ring spot).
Take-all patch	Improve drainage, lower pH if possible, raise HOC, avoid P and K deficiency, avoid using lime.
Yellow patch	Minimize leaf wetness time, avoid excess N, reduce thatch.
Yellow tuft	Avoid excess N, minimize leaf wetness time, improve drainage, mask symptoms with iron.

[a]Adapted from Schumann, 1994.

ing pH, it's important to note that there are instances where this is not only impractical, but nearly impossible (see Chapter 2, Fertility).

Research suggests that trace elements may play a role in a plant's susceptibility to infection. If plants cannot absorb all the nutrients they need, regardless of how obscure the element may be, it is logical that they cannot achieve optimum health, strength, and resistance to infection. A soil test will show the availability of trace elements in the soil, but it is difficult without a leaf tissue analysis to determine if adequate levels are reaching the plant (see Chapter 4, Analysis). The assimilation of nutrients into the plant often depends on the balance of those nutrients in the soil, the pH, and the level of biological activity. Natural acids and enzymes created by soil organisms free trace element ions from parent material or other insoluble sources that can then be absorbed by plants. The optimum pH for trace element availability is ~6.5. In some cases, a pathogen may cause a trace element deficiency, resulting in a better opportunity for infection. The organism that causes take-all patch, for example, metabolizes manganese from the soil, which reduces the amount available to the plant. This deficiency can weaken the plant's resistance to the disease. Supplemental applications of manganese during take-all attack can often reduce its severity.

Stress seems to be the common denominator when diseases (and other pests) are being discussed. Simple practices that can reduce stress—even a little—can often accomplish significant gains in damage control. A researcher at Michigan State University found that light watering of approximately one-tenth of an inch (an average of ~10 minutes of irrigation) each very hot afternoon suppressed most turfgrass diseases. The research suggests that stress is involved in the establishment of an infection, and that mitigating stress by lowering the soil and plant temperatures can ameliorate conditions to the point where natural plant resistance is able to prevent infection. The research found that insect, weed, and thatch problems were also suppressed in the treated plots. Courses with irrigation may have the ability to control many turfgrass pests with about 1/10 inch of water in the mid- to late afternoon throughout the heat and drought season.

Heat and drought take their toll on turfgrass. Both inflict a tremendous amount of stress often resulting in weaker plants that are less able to defend themselves against disease. Many of the most damaging turf pathogens gain significantly better opportunities for infection when the temperature rises. Surprisingly, high air temperature is not as much of a problem as high soil temperature. Turf roots growing in warm soil (60–70° F) can mediate the leaf response to 100° F (38° C) air temperature. However, if the soil is 100°, turf roots are often damaged and no matter how cool the air is, grass quality will decline. This fact may be the main reason why light, frequent irrigation during the hottest part of the day can relieve stress and mitigate infection. Not only does it cool down the plant but the combination of the water temperature and the cooling effect of evaporation reduces soil temperature as well. In hot soils, root growth is inhibited and water uptake is reduced. This in turn affects the water balance between the roots and tops of turf plants. High air temperature also increases the demand for water uptake by the plant's roots and high soil temperature reduces the roots' ability to do so. Consequently, the plant becomes very stressed.

High temperatures reduce photosynthesis (energy production) but increase respiration (energy consumption), resulting in a negative energy balance. The longer the plant consumes more energy than it creates, the weaker and more susceptible it becomes to infection.

Fortunately there are some things the superintendent can do to mitigate high soil temperature.

1. Raise the height of cut. Turf mowed higher has deeper roots, more root mass, and consumes less energy. Additionally, cutting higher will create a denser canopy, shading the soil. This will not only reduce soil temperature but will also decrease the amount of moisture evaporated from the soil's surface.
2. Precondition turf in anticipation of drought. Studies indicate that turf can be conditioned to be more tolerant of drought. Exposing plants to mild drought conditions at low ambient (soil and air) temperatures triggers deeper root growth and more diffusive branching, resulting in better access to water when high heat conditions occur. Infrequent irrigation during cooler periods (i.e., spring and fall) encourages deeper roots and prepares turf for the hot dry season.
3. Use nitrogen judiciously. Many studies have shown that the more nitrogen (beyond optimum) applied to turf, the less root mass is produced. In fact, some studies have

found that the roots of unfertilized turf plots were significantly deeper than the roots in fertilized plots. Plants grown in soils with slight nitrogen deficiencies grow deep roots in search of N. More root hairs are created and the efficiency at which nitrogen is used increases dramatically. We are not suggesting that nitrogen not be used on turf, only that it should not be applied in excess. High concentrations of nitrogen in the soil slow and sometimes arrest turf root development. Excess N will also reduce the amount of carbohydrate reserve in the root system and change the balance of natural hormones that regulate photosynthesis, respiration, and the production of antioxidants. At the same time all of this is going on under the soil's surface, shoot growth is increased, placing an even greater demand on roots for water. Plant tops become more succulent and have thinner cell walls that can be penetrated more easily by pathogens. Some studies show that varying levels of nitrogen have different effects on different cultivars of grass. High levels of nitrogen, for example, caused brown patch suppression in ryegrass but the opposite was true with tall fescue. Other studies show that some diseases such as Bipolaris and Drechslera leaf spot, Pythium diseases, stripe smut, summer patch, and pink and gray snow molds are favored by higher application rates of nitrogen and some, such as stem and crown rusts, anthracnose, red thread, and dollar spot are suppressed. It would not be well advised to prescribe a set amount of nitrogen for every cultivar growing in every situation, but it is much easier to apply more where it is needed than removing some where too much has been applied. There are two schools of thought regarding dormant feeding of nitrogen. Some believe the best time to apply nitrogen is late fall, after the turf has reached dormancy. They believe dormant feeding favors root growth which, in turn, increases photosynthesis and the storage of carbohydrates for the following spring. Others feel that applying nitrogen late in the fall increases the likelihood of winter diseases such as pink and gray snow mold and Pythium snow blight. Where snow mold is a perpetual problem, dormant feeding may not be appropriate.

4. Irrigate lightly and frequently during hot dry periods. About one-tenth of an inch of irrigation will significantly reduce the temperature of both the plant and the soil. Subsequent evaporation will provide further cooling to leaf and soil surfaces.
5. Apply seaweed extracts. The natural hormones in seaweed will help the plant produce more antioxidants, which will reduce stress and sustain a balance between photosynthesis and respiration.
6. Avoid using herbicides, which can inhibit the vigor of turf roots.
7. Encourage biological activity. Humus produced by soil organisms holds more water and stimulates plants to produce finer root hairs. The increased surface area on these roots has greater access to soil moisture.

As mentioned earlier, research has shown that optimum fertility, especially with regard to nitrogen, can aid in the natural suppression of turf diseases. Natural organic nitrogen may be better suited for this task because of its slow release properties and because it encourages the growth of microbe populations that can ultimately compete with, parasitize, or antagonize plant pathogens. Populations of saprophytic and hormone-producing organisms that generate more carbon dioxide for plants and help plants resist stress (respectively) are also encouraged. Clippings contain more than half of the nitrogen applied to turf and it is in probably the most ideal form for turf. If it is impractical to leave clippings behind, then they can be saved, composted, and eventually returned to the turf during topdressing procedures.

Choosing disease resistant cultivars is a wise choice, especially in regions where certain turf diseases are common. Most land grant universities have data on the disease resistance of different seed varieties and new information is continuous. Diversity can also be an important strategy concerning disease control. Most pathogens are relatively specific about which plants they infect. If different varieties exist, the infection of one species may be less noticeable and it may be more difficult for the disease to spread. Research has shown that mixing cultivars of bentgrasses, especially when resistant varieties are included, can improve natural disease control. Indigenous varieties, if available, are sometimes the most disease resistant.

Poa annua (annual bluegrass), although considered a weed by many superintendents, is indigenous in many regions of the country but is most prevalent in the spring. Poa can develop into a good playing surface and, if mowed without the basket, can spread nicely to fill in bare spots on the green. Allowing Poa to become a significant portion of a green is a decision many superintendents face. Often Poa is a sign of other problems such as compaction or mowing too low but there is a case to be made for the use of Poa and other indigenous grass species. Many areas in the country, and probably the world, spawn native bent grasses, paspalum, or other low-growth species that are particularly suited for the region in which they grow. These nameless varieties look good, spread well, and spring back more quickly after pest or environmental damage. Some superintendents cultivate native bent grasses in their nurseries.

Drainage is an important consideration for disease management, one that is usually being made when the green is constructed. Consequently, copious amounts of sand are used to provide what some believe to be a healthy soil environment. Unfortunately, this soil environment cannot support a large and diverse population of soil organisms that others believe is an essential component of a healthy soil. It's a moot point. We are not suggesting that superintendents rototill their greens just to incorporate organic matter. On the other hand, we would advocate a switch from using mostly sand as a topdress material. Using mixes of sand with 20 to 40% topsoil is a common procedure but questions like: where did the topsoil come from; what was the soil analysis; how much organic matter was in it; or how biologically active was it; cannot be easily answered. And what about sand? What is being accomplished with layer upon layer of sand? Thatch reduction? Maybe. Puttability? Perhaps, but at what cost? The surface created by sand topdressing is admittedly smoother but not so much smoother that it is worth sacrificing the natural, biological protection offered by well-aged compost. Many managers believe that drainage must be preserved at all costs or disease will prevail—maybe not. Drainage is good to have but not at the expense of everything else that functions in a healthy soil. Substituting well-made, well-aged compost for some of the sand and most of the topsoil is a compromise that will feed a broad diversity of disease suppressive organisms without jeopardizing drainage. Coauthor Richard Luff, and his father before him, have been using a 50% sand, 50% compost topdress mix since 1965 and drainage has never been a problem.

Dyed-in-the-wool sand users may want to consider another handicap. There's a new disease spreading across the country called Ophiospharella agrostis. This extremely pernicious pathogen attacks both warm- and cool-season grasses but only appears on turf growing in high sand content soil. O. agrostis develops in warm weather and is most commonly found on open, sunny greens. The disease initially looks like a ball mark but as its size increases, it takes on a copper color and has some characteristics of pink snow mold or Fusarium patch. Dead spots have different shapes which makes diagnosis more difficult. The disease leaves crater spots of the greens that take a long time to recover because the pathogen kills the stolons before they can fill in. Fruiting bodies can shoot spores five to six inches into the air, which can then be carried by wind or water. Spores can germinate in one hour in the presence of light and leaf tissue. Damage can begin in less than two hours and the pathogen can infect both leaves and roots. No resistant cultivars have been found but some recover faster than others. Turf growing in soil (not sand) seems to be unaffected by this pathogen.

The importance of good drainage underscores the danger of having a wet environment for a prolonged period of time. The longer turf leaves and the soil surface stay wet, especially in hot, humid weather, the greater the opportunity for pathogen infection. Good air movement is as important as good drainage. Air moving over the grass leaves evaporates moisture and cools the leaf's surfaces. Some superintendents have resorted to using mechanical devices such as fans or blowers to improve air movement over some greens. Research has shown that 10 or more consecutive hours of leaf wetness can be very conducive to disease development such as brown patch and dollar spot. Air movement is most often dictated by Mother Nature but there are a couple of things the superintendent can do to improve the hand that nature deals us. First, take note of the prevailing winds. From what direction do they normally originate? Are there any natural windbreaks such as trees or hedges, between the green(s) and the direction from which the wind blows? If yes, some thinning may be in order. If a building or some other permanent structure is blocking air movement to a green and the green is prone to diseases encouraged by prolonged moisture, mechanical air movement might be a viable option during hot, wet periods of the season.

If overseeding is a regular practice, it might be a good idea to take a look at the turf density. It's common knowledge that a thick healthy stand of turf is the first line of defense against weeds but, taken to its logical extreme, it can create problems too. A stand that is dense enough to inhibit air movement may increase the likelihood of disease infection. Greens grown on soil enriched with compost or high quality organic fertilizers may sometimes grow too thick. Thinning may be advisable before the hot humid days arrive.

Another side effect of overdoing overseeding is the increased likelihood of Pythium disease. Pythium spores sense the chemicals released by germinating seeds. These chemicals act like an alarm clock that wakes up dormant Pythium spores and tells them it's time to go to work. The more seeds there are germinating at once, the louder the alarm. In a biologically active soil, competition for these chemicals is fierce, which reduces the amount available for Pythium spores but planting seed too heavily can make it more difficult for natural biological controls.

Early detection of turfgrass diseases is as valuable to turf health as early detection of cancer or heart disease is to human health but it is difficult for a superintendent to scrutinize the entire course on a daily basis. Ground crew members, however, can be trained to recognize the beginning signs of disease outbreak. It's not important that they be able to identify the pathogen or its stage of growth—only that they are on the lookout for mycelium, discoloration (especially in circular patches), or any other signs of abnormalities. If every member of the crew was required to carry a small note pad and pen with them, they could record where and when the symptom was observed and report it to the superintendent at their earliest opportunity. Some incentives could be offered in hopes of heightening their vigilance. Treatment at earlier stages of development is usually more effective and the disease can often be controlled with less fungicide. Treatments with well-aged compost or compost tea are also more effective at earlier stages of disease development.

Even under the best management practices, some turfgrass loss can occur during prolonged periods of stress. Treatments to the entire course in hopes of preventing a relatively miniscule loss are economically and ecologically impractical. Small losses, especially in problem areas, present opportunities to incorporate disease suppressive compost during renovation. Sometimes it is prudent to give nature an opportunity to resolve the problem on her own before intervening. Often, a combination of natural factors, such as weather changes, will enable turf to quickly repair itself. If overseeding becomes necessary, an application of seaweed extract can hasten the establishment of new seed or sod and also reduce the plants' susceptibility to infection. This relatively brief inconvenience to golfers can provide long-term benefits to the course and the superintendent.

WEEDS

Weeds are plants—often plants with value—that are in the wrong place. In a natural, uncultivated setting there are no weeds. The natural tendency of plants is to use every available space on the soil's surface. If sun, water, and nutrients are available in a space where little or no competition for those resources exists, then there will soon be a thriving plant. Nature provides this opportunity to protect the soil from erosion and to provide the cooling and moisture preserving effects of shade. Many so-called weeds are cultivated and sold as valuable plants for different locations. An expensive installation of a ground cover that creeps into the fairway becomes a weed at the garden's border. Perennials that propagate underground and spread into turf become weeds no matter how rare or beautiful they may be.

What makes a plant a weed is the intolerance of the golfer, the superintendent, or both. Most weeds cannot survive constant and low mowing so there are relatively few that invade turf and significantly fewer that can actually interfere with the game. The problems created by weeds are more aesthetic than they are disruptive to the game. A tree with limbs hanging out over the fairway can influence the course of a golf ball significantly more than a dandelion but neither the golfer nor the superintendent would ever consider using herbicide on the tree because the tree has aesthetic value—the dandelion does not. Depending on the tolerance level of the superintendent, weeds may be much more difficult to control without pesticides than diseases or insects. Zero tolerance on weeds is a tough stance even with an arsenal of herbicides. Part of developing an ecological maintenance program is learning to live with a few weeds.

Most weeds play an important role in the broad scheme of things. Dandelions, for example, are very beneficial plants. Their deep roots return leached nutrients to the surface and are large producers of organic matter. Earthworm populations thrive in their vicinity and the plant does not compete with turf for consequential amounts of nutrients, water, or light. Clover is another beneficial plant that has gained undeserved notoriety. Clover is a legume that can fix free nitrogen from the atmosphere and share it with turfgrass. Clover roots are extensive and contribute a significant amount of resources to soil organisms. These plants are also extremely drought resistant and can stay green long after turfgrass has gone dormant.

There is strong evidence that the incidence of insect predators, parasites, and antagonists increase where there is a greater diversity of plants. Some weeds can act as bait plants, drawing herbivorous pests away from turf. Plant diversity is an important ecological strategy. Depending on only one species of plant to survive and thrive in a given area is too high a risk for Mother Nature. The diversity of a natural setting not only provides better survivability but also creates a habitat for a more balanced ecosystem. Only one variety of plant in a given habitat can limit the diversity of other organisms, which may also limit the natural protections they offer. In terms of weed control, a diverse mixture of grasses is often more competitive than a single cultivar would be. If grasses are to dominate an area without the use of herbicides, they must adapt well to environmental conditions. Given the diversity of conditions that exists in most settings, a successful strategy may be to cultivate a variety of different grasses where at least one or two cultivars will thrive if others fail. Diverse combinations of turf seeds are often a reliable strategy for maintaining a dense sod in spite of nature's unpredictable conditions. Whenever a cultivated plant does not adapt well to existing conditions, Mother Nature will provide a substitute. Unfortunately, the plant that nature chooses is often not acceptable to the superintendent. Diversity on the fairway is easier to tolerate than on the greens. However, researchers recently found that a mix of different bentgrass varieties has better resilience to stressful conditions.

Poa annua (annual bluegrass) can be a problem weed for many superintendents but others tolerate or even welcome its presence. Poa can withstand compaction better than bentgrasses and its appearance may be a sign that the green needs conditioning but, like bentgrass, it tolerates low cutting heights and, if mowed with the basket off, will spread by seed into thin areas of the green. This fill-in capability is a lifesaver for some superintendents. Stands of Poa, especially where its coverage is expanding, may also be a sign of poor drainage, overirrigation, or excess nitrogen.

There are two different subspecies of Poa, the annual variety (*Poa annua* ssp. *annua*) and a perennial variety (*Poa annua* ssp. *reptans*). Differences between the two varieties include root system, growth patterns, seed production, and life cycle. The annual variety has a shallow root system, grows erect, produces seed heads in May and June, and has only one life cycle per season. The perennial variety has a strong, fibrous root system, grows prostrate, produces seed several times per season, has a life cycle that can last several seasons, and is more difficult to control. Some superintendents have successfully controlled *Poa annua* ssp. *annua* by allowing the affected green to go dormant for a short period during the hot summer months. Bentgrass will recover from a brief period of dormancy but the annual variety of Poa will, almost always, die off. Unfortunately, if Poa occupies a conspicuous percentage of the green, the end result of this strategy is more unsightly than the weed. *Poa annua* ssp. *reptans* cannot be controlled in the same way. Conditions that favor Poa over bentgrass include compaction, excess nitrogen, low height of cut, uncollected clippings, and excessive irrigation. The lower the superintendent needs to mow his greens, the more likely he will have a problem with Poa, but it is only a problem if he can't accept Poa as a suitable playing surface.

Soil conditions have a profound effect on the strength and competitive abilities of turf. Many weeds that exist in turf are often a result of poor soil conditions, such as compaction, infertility, improper pH, or excessive porosity that do not favor the preferred turf cultivars. Intervention at the soil level can often provide a solution that is not only effective but long lasting as well. Many agronomists have observed that the proper balance between calcium and magnesium can significantly reduce the need for herbicides. The relationship between this balance and the rollback of weeds is not well understood but balanced fertility offers both plants and support organisms optimum conditions for competitive performance. Observations indicate that soils with a low calcium to magnesium ratio (below 7:1) are more prone to compaction which favors weeds that can tolerate heavier soil density (see Chapter 4, Analysis).

Herbicides may eradicate weeds, but they do not change the conditions that gave weeds their competitive edge. Additionally, some herbicides can increase pest insect activity by suppressing predators, stressing plants, or both. Recent research from Cornell shows that applications of certain herbicides can also increase the severity of some disease symptoms.

An increase in populations of herbivorous insects can stress turf to the point where weeds are more easily established. Researchers discovered that all living things give off radiation in specific wavelengths (Callahan, 1975). When less than ideal conditions cause problems for plants, those wavelengths are altered and may act as signals to herbivorous insects. The insect can, in turn, create conditions that favor one species of plant over another. In a situation where turf is subclinically ill (i.e., plants that have a health problem but are not as yet showing symptoms), the plants' altered wavelengths may signal insects that then cause further stress to the turf. Insect activity can suppress the health and vigor of turf and favor the establishment of weeds. This phenomenon can be a double-edged sword. Weeds that are subclinically ill due to the unyielding competition from healthy turf may signal insects to act as a selective, biological herbicide. The general message nature gives us is to strengthen the preferred species rather than weaken the unwanted. Mowing with a dull blade, for example, may exert just enough stress on a stand of turf to trigger either a disease or insect problem which could, ultimately, create the ideal setting for weed infestation.

The unforeseen circumstances that thin out an area of turf are sometimes unavoidable. Vigilant attention to injured areas, however, can mitigate problems without resorting to the use of herbicides. If these areas are prepped for new seed as soon as possible, the encroachment of weeds may not occur. Plant and weed residues should be removed to avoid the possible production of allelopathic chemicals that could inhibit the germination of new seed. Some superintendents create a mixture of seed and aged compost, which they rake and tamp into injured areas. This method is very successful; however, compost that is not well cured can have a high salt level, which may kill germinating seeds. If one employs this method, it is a good idea to have a maturity test performed on the compost before use (see Chapter 3, Compost). Generally, a conductivity value above 4 millimhos per centimeter (0.4 siemens per meter) is too salty for seed.

If all else fails and it becomes necessary to use a herbicide it is wise to investigate the tolerance different species of grasses have to the product one intends to use. If the use of a herbicide further weakens the turf we are trying to protect, then the apparent purpose of the product is defeated. In addition to weakening the turf, every herbicide ever tested has effects on nontarget organisms. Qualifying or quantifying these effects would be difficult in light of the other factors that influence the diversity and quantity of soil and leaf surface organisms. Tests indicate, however, that natural disease and insect suppression is often inhibited after one or more applications of certain herbicides. Healthy populations of soil organisms are responsible for numerous other benefits that indirectly affect weed growth. These benefits generally favor the turf which can then be more competitive and successful. Soluble seaweed extract, humic extracts, or combinations of both can be applied with most herbicides to assuage any possible impact the chemical might have on turf and soil organisms. These materials will complement the application of just about any pesticide but a physical compatibility test is always recommended first.

Often, effective weed control in turf is as simple as raising the height of cut. This action enables the turf to be more competitive by using the sun's energy for itself and blocking it from fueling unwanted species of plants. For every eighth of an inch that a mower is raised, there is a 30% increase in leaf surface area. That increase causes a relative increase in photosynthesis, which feeds a larger and healthier root system and a larger, healthier population of organisms in the rhizosphere. The lateral roots (rhizomes, stolons, or tillers) of some turf varieties often develop new plants that further thicken the stand, providing even more competition. Spring is an important time to mow higher, as the plant is in its reproductive stage, creating new tillers, stolons, and rhizomes. Raising the height of cut in the spring can control many weeds all season long. Turf roots can produce allelopathic chemicals capable of suppressing weed seed germination. The greater the root system, the more allelochemicals can be produced. Additionally, the lush growth typically observed in spring is the plant's way of processing nutrients into proteins, carbohydrates, and other compounds via photosynthesis. Mowing at an extremely low height of cut (HOC) inhibits this process and can lead to a host of other problems.

HOC experiments have demonstrated that grass mowed higher can suppress many weeds, including crabgrass, annual bluegrass, goosegrass, and dandelions. It is obviously impractical to mow at heights that impede the game. On the other hand, mowing stressed turf at an unreasonably low HOC is an open invitation for weeds, especially annual bluegrass.

Mowing lower, however, can also be a useful strategy for controlling some already established annual weeds but timing is important. The height of cut needs to be lowered and the clippings collected when seed heads have formed but seeds are still attached. If timed correctly, the amount of viable seed can be significantly reduced but the HOC should not be lowered too much. Stressing the turf by cutting too low can give other weeds a distinct advantage. Weeds effectively controlled with this strategy include annual bluegrass, crabgrass, goosegrass, foxtail, barnyardgrass, fall panicum, and dallisgrass. A single weed plant can produce from 1,000 to 500,000 seeds and under the right conditions, many of those seeds can become more seed producers.

Identifying weeds is as important as recognizing insect pests or plant pathogens, especially if the plan is to use herbicides. Conventional controls work best if they are treating the pest for which they were designed. Identity is also important because there are conditions in which certain weeds thrive. For example, crabgrass seems to like sandy, well-drained soil that can send cultivated grasses into early dormancy. Dandelions can indicate soil compaction, low pH, or both. Attention paid to all the different stories that weeds tell is a valuable approach to controlling them but, in a situation where many different weeds are telling many different stories, cultural controls may be difficult to prescribe. If symptoms such as weeds are indicating problems with soil fertility, then a soil test may be an easier way to pinpoint those problems (see Chapter 4, Analysis). Soil tests, however, will not expose physical problems, such as compaction, drought, or excessive moisture. Some of these conditions would be recognized while drawing samples from the area. Most labs do not test for available nitrogen unless one specifically asks for it so a soil analysis may not indicate a condition of excessive N but a leaf tissue analysis would. When a human patient exhibits classic symptoms of a certain condition, a doctor will usually perform a test to confirm the condition before prescribing medication. In turf, weeds may be indicating a specific condition such as infertility or incorrect pH, but a soil test, leaf tissue analysis, or both will confirm the diagnosis. It is helpful to understand weeds and in what conditions they thrive (see Table 5.2), but it is more important to analyze the soil as a first step in developing a weed control plan. Biological respiration and bioassay tests are other tools that can help the superintendent grow stronger, more competitive turf. There is a balance of soil organisms that is as important

Table 5.2. Common Conditions that Promote Certain Weeds[a]

Weed	Conditions
Algae	Excessive surface moisture
Annual Bluegrass	Excessive surface moisture, Compaction, Mowed too low
Barnyardgrass	Poor drainage
Birdsfoot Trefoil	Droughty conditions, Low nitrogen
Black Medic	Droughty conditions, Low nitrogen
Broadleaf Plantain	High pH, Compaction
Buttercups	Poor drainage
Chickweeds	Mowed too low
Cinquefoil Species	Droughty conditions, Excessive surface moisture, Low pH, General low fertility
Clover Species	Low nitrogen
Coltsfoot	Poor drainage, Low pH
Common Chickweed	Too shady
Common Mullein	Low pH, General low fertility
Corn Chamomile	Poor drainage, High pH
Corn Speedwell	Compaction
Crabgrass	Droughty conditions
Creeping Bentgrass	Poor drainage, Excessive surface moisture, Mowed too low
Creeping Speedwell	Too shady
Creeping Thyme	High pH
Curly Dock	Droughty conditions
Docks	Poor drainage, Low pH
English Daisy	Low pH
Foxtail Species	General low fertility
Goosegrass	Droughty conditions, Compaction
Hawkweeds	Low pH, General low fertility
Henbit	General low fertility
Hop Clover	High pH
Knawel	Low pH
Lady's Thumb	Poor drainage, Low pH
Leafy Spurge	Droughty conditions
Mallow	General low fertility
Moss	Excessive surface moisture, Mowed too low, Too shady
Mouse-ear Chickweed	Too shady
Nutsedge	Poor drainage
Pigweed	Droughty conditions
Pineapple Weed	Compaction
Plantains	Poor drainage, Mowed too low
Prostrate Knotweed	Compaction
Prostrate Spurge	Droughty conditions, Compaction
Rabbit Foot Clover	Droughty conditions, Low pH, High pH
Sheep or Red Sorrel	Low pH
Speedwell	Droughty conditions, Mowed too low
Vetch Species	Low nitrogen
Wild Carrot	High pH, General low fertility
Wild Parsnip	General low fertility
Wild Radish	General low fertility
Wild Strawberry	Low pH
Yarrow	Droughty conditions
Yellow Woodsorrel	Droughty conditions

[a]Adapted from Bosworth.

to turf performance as the correct balance of soil nutrients. Although a book is generally less ephemeral than a small business, it is worth mentioning a lab in Oregon that performs bioassay tests and makes recommendations to establish and maintain optimum balances of organisms on a golf course (Soil Food Web).

Much has been published over the years that identifies various conditions in which certain weeds tend to thrive. This information gives us hints as to what may be causing a weed problem. It makes sense, however, to first look at the conditions in which the desired species of grasses thrive, because optimal conditions for turf will limit weed invasion more than anything else. Balanced fertility, correct pH, good soil structure and porosity, ample amounts of organic matter, adequate level and balance of biological activity, and proper maintenance procedures are the factors that influence the aggressive nature of turf. If all of these conditions are initially addressed and there is still a weed problem, then, at this point, the weed information may become very valuable. Most of the conditions that favor weeds do not favor turf, however, so if ideal conditions for turf have already been created, chances are that the conditions that favor weeds have already been addressed.

The broad use of preemergent herbicides is well accepted by many turf managers who do not actually know that it is necessary. One reason may be paranoia—the fear of losing one's job should weeds rear their ugly little heads, especially just before a tournament. The expense of using preemergent herbicides should be considered even if the ecological factors are ignored. Using preemergent herbicides is analogous to a person taking antibiotics to protect himself or herself against a bacterial infection. The outcome for this person, however, would be a weakened system that is more susceptible to infection. Preemergent herbicides are designed to suppress the germination of weed seeds while having no effect on nontarget organisms. Ecologically speaking, it is not logical to assume that dose after dose of these products has no impact on the natural system. Unfortunately, if any part of the natural system is stressed, the resulting chain reaction may eventually result in other problems. Tests from Cornell show increased severity of disease symptoms when herbicides and insecticides are used, which suggests that the pesticides are adversely affecting disease suppressive organisms, plant defenses, or both.

Preemergent herbicides are sometimes used for a much longer period of time than they are needed, especially for controlling crabgrass. Preemergents, by nature, allow seeds to germinate but not grow. The resulting depletion in the seed bank lessens the need for preemergents unless some crabgrass plants succeed in growing and setting seed. It is reasonable to assume that after a few years of annual preemergent applications, the seed bank has been sufficiently depleted and there is no immediate need for preemergent applications. The superintendent, however, will never really know if he continues with the annual (or more frequent) tradition of applying preemergent herbicide. In the meantime, it is likely that repeated applications of herbicide are causing some impact on soil organisms and turf plants. Although difficult to quantify or qualify, it is likely that the impact reduces the opportunity and the ability for turf to compete as a healthy member of the ecological community.

Weeds such as crabgrass are opportunistic but have a difficult time getting established if the right conditions do not present themselves. First and foremost, crabgrass needs open space. Experiments at Cornell University confirmed that the larger the open area, the more likely it is that crabgrass or goosegrass will become established. They also discovered that a small amount of thatch was a deterrent to crabgrass germination. This research suggests that paying close attention to divots on the course can significantly reduce the opportunity for crabgrass or goosegrass establishment. Vigilant applications of a divot mix that quickly reestablishes turf can reduce the need for herbicide. Divot mixes can vary considerably but it is advisable that well-aged compost be a component. Compost will retain more moisture and contains organisms and compounds that can not only increase the germination time and percentage, but also protect the seedling from pathogens such as Pythium. It is important that the compost be well-aged and well-made. Immature compost may contain salts that can suppress or kill germinating seeds. If seed is used in the divot mix, only those within one-half inch of the surface will probably germinate and survive. Some superintendents use clippings, sand, and compost (mixed in equal portions) as a divot mix with surprisingly good results. It looks green, and adds nitrogen, organic matter, moisture, and possibly some Poa seed. The mix fills in remarkably faster than nothing.

Another aid to controlling some weeds is grass clippings management. To collect or not to collect is the question and the answer may significantly reduce one's need for herbicide. Depending on conditions, the decision to collect clippings may encourage or discourage weeds. Clippings from many different species of turf plants are thought to contain allelopathic compounds that suppress the germination and/or growth of certain weeds. Furthermore, the distribution of clippings can act as temporary mulch, prior to decomposition, that not only suppresses weeds but also provides nutrients and preserves moisture for the established turf. Fifty-eight percent of the nitrogen from turf fertilizers resides in the clippings. Redistribution of that nitrogen and the other nutrients in clippings gives turf fuel to compete even more aggressively with weeds. The benefits of leaving the clippings behind include improved water infiltration, higher populations of earthworms, greater root production, and disease suppression. All these benefits may have either a direct or indirect influence on the suppression of weeds. In instances where excess nitrogen is favoring weeds like annual bluegrass, crabgrass, foxtail, barnyard grass, and fall panicum, or weed seed heads are visible, clippings can be collected to reduce available N and limit seed dispersion. But the value of the clippings should not be ignored. Clippings are an important ingredient for the compost pile (see Chapter 3, Compost). Annual bluegrass seed dispersion can be significantly reduced by collecting clippings but some superintendents value its ability to fill in bare spots, and mow with the basket off to better disperse the seed.

Nitrogen applied insufficiently or to excess can favor some weeds directly and others indirectly. Crabgrass and annual bluegrass are both more competitive when levels of available nitrogen are above optimum, whereas clover and other legume weeds are favored by low levels of available nitrogen. Excess nitrogen that encourages disease outbreak indirectly favors many opportunistic weeds that become established in the bare spots that remain. The same is true where excess nitrogen in plant tissue makes turf shoots, roots, or both more appetizing to herbivorous insects or larvae. Inadequate nitrogen can also trigger insect problems because the plant is weaker and less able to produce defense compounds. Anything that thins turf also gives weeds an opportunity to move in. Knowing the nitrogen needs of each type of turf on the course is an important step toward preventing the encroachment of weeds, not to mention pathogen infections and insect invasions. Biological regulation of N can synchronize plant needs with availability (see Chapter 2, Fertility) but resources that support soil organisms must be present.

Reseeding areas in the fall instead of the spring is another practical way of controlling many weeds. Competition from weeds is usually greater in the spring than in the fall when most annual weeds are beginning to die off. There is evidence that when seeds germinate, they give themselves a competitive edge by releasing natural plant hormones that inhibit the germination of other seeds in their immediate surroundings. This might be as true for weed seeds as it is for grass seeds. This factor makes fall planting, when the germination of weed seed is at a low ebb, more sensible. In many regions of the U.S. the reliability of rain is greater in the fall than in the spring. High spring mortality of grass seeds is something most superintendents have experienced and the replacement seed never seems to do as well, possibly because of the natural hormones released by other germinating seeds. Bentgrass is an exception to this strategy. Its best chance for successful establishment is generally in the summer or early fall months when soil temperatures are higher.

Another important method of weed control is the proper selection of seed varieties for fairways, tees, and greens. Varieties that struggle to survive under unsuitable conditions are less able to successfully compete with weeds. A reiteration of diversity is appropriate here. Conditions such as sun and shade may be easily identified, but some invisible conditions may limit the success of a select species of grass. Diversity in a mix can offer strength over a broader range of conditions and increase the odds of producing a dense sod capable of crowding out most weeds.

The soil inherently contains millions of weed seeds deposited by wind, water, and animals. These seeds can stay dormant for hundreds of years until, somehow, they reach the proper soil depth where warmth, moisture, and other factors will trigger their germination mechanisms. Deep tillage during turf renovation can bring many of these seeds to the surface where they can germinate. Shallow soil preparation may be a better alternative if the creation of a seedbed is all that is necessary. Seeding machines that plant without turning the soil help to reduce the number of dormant seeds that can germinate when the soil is disturbed. If it is necessary to burn down existing foliage before overseeding, there are effective contact herbicides with relatively benign active ingredients. Some are made with citric acid (St. Gabriel Labs) and some with natural fatty acids (Mycogen). This leads to the subject of regular overseeding.

Senescence is a term used to describe the natural aging process of all living things. The shoots of turf plants live an average of about six weeks and then die off. Fortunately, new shoots are being produced from the crown to replace the dead ones and it is difficult to actually observe senescence in turf. Regular applications of seaweed extract can retard senescence and can help maintain stand density and suppress weeds. Theoretically, the perennial grass plant is supposed to generate new growth from its crown year after year, ad infinitum. In many varieties it will even generate new crowns. Unfortunately, there are other factors in nature that have a tendency to interrupt this tradition and cause permanent senescence in turf. Overseeding fairways, tees, or greens can sometimes inhibit senescence and thwart the typical invasion of weeds in a declining turf. The infusion of new seed into existing turf is, in essence, an injection of youth into a natural aging process. If a parcel of turf is in natural decline, it is eventually replaced by nature's choice of seed. If overseeded, a more preferable variety is selected to accomplish the same end. This philosophy is practiced successfully by some superintendents. Overseeding greens is a more challenging proposal. The most difficult part is making the transition from the existing surface invisible to the golfer. This is relatively easy if the superintendent's goal is an annual bluegrass green. Mowing with the basket off dispenses the seed quite effectively. Increasing the percentage of bentgrass on a green, however, can be more difficult. Before overseeded bentgrass can be successful, the conditions that favor its existence over Poa or other weeds need to be established. Successfully overseeding Poa greens with bentgrass is more difficult to hide from the golfer. If the transition needs to occur in a short period of time, Poa needs to be noticeably stressed before bentgrass can succeed. This procedure is most often used to repair large damaged areas that can't repair themselves in a reasonable amount of time. If, however, the superintendent can make the transition slowly, over a period of several years, it is possible. Conditions that favor Poa over bentgrass include low HOC, superfluous nitrogen, and compaction. Changing these conditions and overseeding can slowly increase the percentage of bentgrasses to acceptable levels.

As with anything else, overseeding can be done to excess. If seed is applied to excess, it can become more prone to certain diseases, especially those that thrive in environments where moisture is retained for long periods of time and air entrainment is inhibited. Pythium root rot can also be a concern because of an increase in chemical signals that activate Pythium spores. These signals come from germinating seeds and the more seeds that are germinating, the greater the signal. Common sense should intervene when one is about to overseed an area that could not look any better than it already does.

There are commercially available organic herbicides. These are weed control products that are biologically or botanically derived, exhibit little or no environmental persistence, and do not inhibit soil biological activity. Some of them (e.g., corn gluten) actually stimulate and feed biological activity to the benefit of grass plants. Corn gluten is a by-product of the corn syrup industry that releases allelopathic chemicals as it decomposes in the soil. These chemicals act as a preemergent herbicide by suppressing the growth of roots on germinating seeds. Because it is a food protein, it also provides nitrogen to the existing turf. In some cases, the amount of corn gluten needed to control weeds will deliver too much nitrogen. So although the level of weed control may be excellent, the potential for disease to develop is increased. Experiments with corn gluten suggest that the more biologically active a soil is, the more effective corn gluten will be, and that less of it is needed. Smaller application rates reduce nitrogen contributions and the possibility for disease outbreak. The cost of using corn gluten as a weed control is also significantly reduced. Corn gluten contains 10% nitrogen. Fertility programs that include nitrogen need to be adjusted when corn gluten is used so that excessive N is not applied. Corn gluten—labeled as a herbicide—is commercially available under a number of different brands.

Scientists have isolated five of the dipeptides identified as herbicidal compounds in corn gluten and found that one, alaninyl-alanine (endearingly referred to as Ala-Ala) had the greatest impact on root growth. They found that small amounts of this dipeptide had numerous herbicidal effects on the roots of germinating ryegrass plants. This research may, at some point in the future, produce a corn gluten derivative that can effectively control germinating weed seeds without adding excessive nitrogen.

As innocuous as some of these new tools may seem, they still address only a symptom, not the problem. They do not correct the conditions that encouraged weeds to grow in the first place. The development of these products is analogous to the introduction of methadone to heroine addicts. The addiction hasn't changed, only the substance of abuse. Using more environmentally benign materials is always preferred, but is not necessarily the answer to the problem. If a product like corn gluten is used as a tool to accomplish the objective of soil and turf improvement to a point where herbicide is no longer needed, it becomes extremely beneficial. However, if it is used only to eliminate symptoms on a regular basis, it can eventually begin to cause other problems.

Regardless how a soil looks or feels, it needs to be tested and perhaps treated to obtain balanced fertility and active biology. Beginning with a well-balanced and fertile soil environment will do more to prevent weed problems than any arsenal of herbicides. Cornell University often prefaces their weed control information with the statement that *"The first line of defense against weeds is a dense sod."* To obtain a dense sod one must provide the necessary conditions.

There are many conditions, either physical, chemical, biological, or combinations of the three, that favor the proliferation of weeds over turf. Table 5.2 shows some of the conditions that can give weeds a competitive edge but the occurrence of just a small number of weeds or different varieties of weeds in the same place may indicate a problem other than what is listed. Sometimes weeds may grow well in conditions that are opposite of what the table may indicate just to prove there are exceptions to every rule. One rule, however, seems to apply to every situation: weeds tend to proliferate where turfgrass won't. Conditions that stress turf plants generally favor weeds. Any time the growth of a grass plant is inhibited by stress, another species of plant can gain a competitive edge. If fertility is the problem, testing the soil can usually discover it and administering the proper inputs can correct the imbalance but timing is also an issue. Common sense suggests that the best time to feed turf (or any other living thing) is when it is hungry. Turf, like most other living things, will have the greatest appetite when it is expending the most energy, i.e., when it is growing. Fertilizing during periods of the season when turf growth is waning may be supplying nutrients to other, less desirable plant species. If a soil analysis indicates an adequate level and balance of nutrients in the soil, care must be taken to give plants only what they need. Phosphorus (P), for example, is a nutrient that is very important to germinating seeds but established turf rarely needs much more than what is already in the soil. Applications of P in excess of what turf needs may only serve to increase the germination of weed seeds.

Low levels of organic matter are not as easy to correct, but can be addressed by different methods. These include mowing higher for greater root development and applying well-made, well-aged compost as part of a topdress mixture. Organic matter plays a major role in establishing and maintaining a dense turf. It is home to billions of soil organisms that perform hundreds (if not thousands) of beneficial functions including disease and insect suppression. Bare or thin spots created by insects or pathogens are susceptible to weed encroachment. Soil organisms are also responsible for making many mineral nutrients available to plants. Growing and maintaining a dense turf without adequate organic matter requires more intensive management (see Chapter 6, Cultural Practices).

Compaction gives many weeds an advantage but core aeration should only be done during periods of the season when turf is growing vigorously. On the other hand, aeration, verticutting, brushing, or other activities that open the turf canopy should be avoided during times of the season that are ideal for weed seed germination. It's impossible to escape bringing some seeds to the surface, but if conditions are less than ideal for germination, the established turf has a distinct advantage. Harvesting cores can reduce the amount of weed seed that is spread over the soil's surface. Core aeration during dormant or slow-growth periods (e.g., summer), increases the potential for weed encroachment.

Sometimes nature delivers superfluous water and, aside from an underground drainage system, there is not a lot the superintendent can do about it. Redirecting traffic and vehicles away from areas where the soil tends to stay saturated longer can prevent

The core harvester collects cores which are later added to compost piles. Harvesting cores can reduce the number of dormant weed seeds brought to the surface by cultivation.

compaction that can often result in an outbreak of weeds that tolerate a heavier soil density.

Stress caused by heat and/or drought is more difficult to control. Mowing higher can help because it not only shades the soil, preserving moisture and lowering the soil surface temperature, but a higher HOC also encourages deeper rooting and greater access to more soil moisture. Irrigation can mitigate the effects of heat and drought but irrigation techniques are extremely important (see Chapter 6, Cultural Practices). The last thing turf needs when it is already stressed by heat and drought is to be drowned. Overirrigation forces the oxygen out of the soil and can suffocate the roots. It can also make the soil more prone to compaction. In many cases, light, frequent watering (~0.1 in./d) during the most stressful part of the day can help maintain turf density. Deep watering may also be necessary from time to time but saturating the soil should be avoided.

In the spring, some of us get a false sense of financial stability—until we remember that Uncle Sam hasn't been paid yet. That same feeling of euphoria, followed by frustration, comes when summer heat and drought hammer the spring turf that the ground crews have maintained so beautifully. The stress from high temperatures and low levels of moisture can change a thick, healthy turf into one peppered with problems. The turf's ability to compete with weeds, resist disease, and recover from insect damage are all but gone in just a few short weeks. Aside from irrigation, it seems as though there is little recourse except to stand by and watch its decline.

Fortunately, there are a few things we can do to combat the summer's impairment of turf. First and foremost is raising the HOC. Taller grass can shade the soil, significantly reducing the amount of heat the soil absorbs from the sun and the amount of moisture evaporated from the soil surface. Taller plants grow deeper roots that have both better access to water and better storage capabilities. Water use is actually increased slightly from the greater amount of leaf surface area on taller plants. The physical process of evapotranspiration through the leaves, however, can cool the ambient temperature by

7–14° (F). The longer the grass can stay green, the longer the cooling mechanism can remain working. The slightly higher water consumption is more than compensated for by increased access to soil moisture from a deeper and more diffusive root system.

Seaweed extracts and humic biostimulants are inexpensive materials that can also mitigate the effects of stress. Research has discovered that the natural growth regulators in seaweed products can give turf significantly greater drought resistance and stimulate greater root development. The greater resistance turf has to stress, the better it can compete with weeds.

VAM is an acronym for Vesicular Arbuscular Mycorrhizie. This is a family of beneficial fungi that colonize the roots of perennial plants and, on the plant's behalf, reach into the soil's depths for moisture and mineral nutrients. All they ask in return is a small amount of the photosynthesized carbohydrate that is released through the roots. This symbiotic relationship exists in over 80% of all grass species, and can increase the plants' heat and drought resistance significantly. There are commercial preparations of VAM and other mycorrhizal inoculants available that can colonize the roots of turf plants but they are best applied when seed is initially set. These inoculants must make contact with the plant roots in order to begin colonization and they are not able to move through the soil. Core aeration presents an opportunity to inoculate roots but if turf is grown in a healthy, biologically active soil, chances are good that natural inoculation will eventually take place. Mycorrhizae have been around for over 400,000,000 years but are very sensitive to soil disturbances. They are ubiquitous in forests and other areas where the soil is largely undisturbed but adequate populations do not exist in urban and suburban environments where the soil has been moved, removed, shifted, turned, relocated, piled, and compacted. Their numbers are also impoverished on agricultural lands that have been plowed, harrowed, tilled, and treated with chemical fertilizers and pesticides. These are the environments where golf courses are often located so natural inoculation is more difficult, especially if an abundance of chemical pesticides and fertilizers are used. Mycorrhizal inoculants are most successful on turf roots if they have been applied at planting but can also succeed if they are worked down into the thatch layer during topdressing operations. Commercially available inoculants will have specific instructions but, generally, 30,000 to 70,000 spores per 1,000 ft^2 are necessary for good inoculation. Unfortunately, if fungicides are used, that relationship may be compromised or eliminated. Combating disease with well-made, well-aged compost, compost tea, or a program of building soil organic matter can not only provide long-term disease protection, but will encourage the useful relationships VAM have with turf. Mycorrhizae fungi are aerobic organisms and need oxygen to live. As the soil becomes compacted and oxygen levels are reduced, so are the symbiotic abilities of VAM. Once this occurs, they will form dormant spores on the roots and wait for conditions to improve before they become active again. Mycorrhizae can form a protective layer around the roots and prevent pathogen spores, such as Pythium, from becoming established. Pythium and other root diseases indirectly encourage weeds by creating openings in the turf canopy. VAM indirectly suppress weeds by improving drought resistance and resilience from other environmental stresses that inhibit turf's ability to maintain a dense canopy.

Another nonaction we can take to alleviate heat and drought stress is to stop collecting clippings. Clippings returned to the turf can provide several benefits. Between the time that they fall and decompose, they act as temporary mulch that can lower the soil surface temperature. As moisture evaporates from the fallen clippings, the surrounding air is cooled and stress from heat is further mitigated. As mentioned before, clippings can produce allelochemicals as they decompose. Some researchers believe that these allelopathic chemicals can suppress some crabgrass germination. The nitrogen and other nutrients inherent in clippings is a perfect slow-release fertilizer that can help grass stay vital during stressful periods. Additionally, as clippings decompose, valuable carbon dioxide is generated for grass plants. Clippings left behind have also shown disease suppressive properties which can offer some protection during the time of year when plants are more susceptible to infection. Bare spots created by pathogens are open invitations for weeds to become established. Clippings should be collected if there is any sign of disease. Mowing greens with the basket off needs to be done when leaf surfaces are dry and if any clumps form, they should be poled (see Chapter 6, Cultural Practices). Clippings encourage earthworm activity, which not only increases the soil's water holding capacity but, at the same time, reduces thatch. Pesticides that suppress earthworms are inadvertently contributing to thatch and not helping the drought resistance of the turf. Excessive thatch can decrease the turf's resistance to drought because it holds very little moisture and encourages shallow roots. A small amount of thatch can be beneficial to turf. It can mitigate stress from traffic, lessen the impact of heavy rain, reduce direct sunlight to the soil's surface, and help exchange gases with the atmosphere. A thin thatch layer can also suppress the germination of many annual weed seeds because the seeds cannot make adequate contact with the soil.

Irrigation, if available, can be the ultimate treatment for stress caused by heat and drought. The misuse of this tool, however, can do more harm than good. Researchers disagree on the ideal frequency and duration of irrigation. Some believe that deep and infrequent watering is best, while others feel that high frequency, low volume irrigation is optimal. Very little research has been done on irrigation frequency and volume. Some research suggests that light watering (one-tenth of an inch) on a daily basis is best for relieving heat and drought stress. Common sense dictates that there is no single irrigation program that ideally fits every course under every condition. Deep sandy soils with little water holding capacity may be better served with low volume, high frequency irrigation. But this would probably be inappropriate for heavy soils that do not dry out as quickly.

Many superintendents irrigate based on evapotranspiration calculations, i.e., the amount of water that they apply matches, as closely as possible, the amount of water lost from both evaporation from the soil and transpiration from plant leaves (see Chapter 6, Cultural Practices). Maintaining this balance seems to provide turf with the ideal amount of moisture which, in turn, maintains its competitive characteristics throughout the hot and dry season. Replacing only the amount of moisture that is lost from plant transpiration and soil surface evaporation is often ideal and can eliminate the often-costly error of excessive or insufficient irrigation.

Using a moisture meter to monitor soil water is a good idea, but the optimum amount of moisture varies in different types of soil (see Figure 5.3). If moisture meters are used, it is necessary to analyze the texture of a soil to determine its field capacity (the volumetric percentage of moisture that fills half of the soil's pore space). If a superintendent works with many different types of soils on a given golf course, it may be expensive and laborious to do a texture analysis of each different type of soil. This information, however, is very valuable if the overall goal is to maintain a dense turf canopy. (See Chapter 4, Analysis for an inexpensive on-site method of texture analysis.)

Biological weed control is being researched with increasing vigilance. Different pathogens that infect only the target plant (weed) are being isolated and tested, some with very promising results. Most of these organisms, however, are facultative, which

FIGURE 5.3. SOIL MOISTURE RELATIONSHIPS

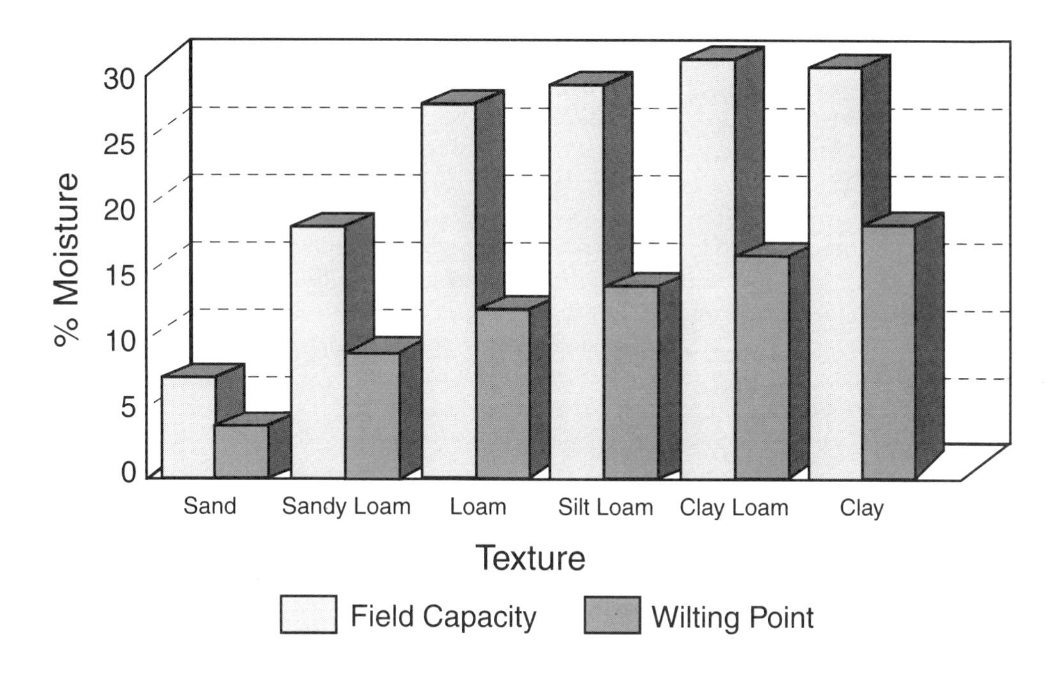

means that infection relies on a specific set of conditions. Those conditions may vary with each pathogen and might include temperature, moisture, compaction, fertility, pH, HOC, and competition from other soil organisms. Since one of the main objectives in an ecological golf course program is to maintain a high level of biological activity and reduce conditions that stress turf, the effectiveness of biological weed control inoculants may be very low. If the superintendent has created conditions that are hostile to plant pathogens, then introduced pathogens may have the same difficulty getting established. Researchers at the University of Guelph, instead of isolating and breeding a specific pathogen, have discovered some of the foods that enable certain biological weed control agents to flourish. They found that adding durum semolina, guar gum, and gluten flour increased the effectiveness of indigenous pathogens that attacked dandelions (Neumann and Boland, 1999).

Algae are not really weeds but can be a problem on golf greens, where the turf is mowed very close and irrigation is at a maximum for maintaining health and color during the hot, dry months of the year. Algae usually occur where there is excessive moisture, high fertility, and low competition from turfgrass. Algae are photosynthesizing organisms and like full sun, but can also exist in shady areas. Controlling algae ecologically is similar to managing weeds. A well-maintained, thick turf is the best defense against algae. Often algae begin in areas of poor drainage. Chronically moist conditions can stress turf and lower its ability to compete for space and sunlight. Algae, on the other hand, thrive in excessively wet areas. Problem areas may require drainage tiles as a permanent solution. Overirrigation may be another cause of algae encroachment.

Too much water can stress grass and provide favorable conditions for algae. Fertilizers applied in excess (especially acidulating fertilizers such as ammonium sulfate) can also encourage algae. Additionally, fertilizers with high salt content can stress turf, whereas algae can tolerate higher levels of nutrients. Mowing higher can protect against algae, which tends to grow in thin layers close to the ground.

Eradication of established algae colonies requires either physical removal or some type of algicide. Household products such as Lysol disinfectant or Clorox bleach with a surfactant are very effective at killing algae, but can create problems. The first is that these products are phytotoxic—they can kill the grass if the dose is too large. This may not be a big problem if there isn't much grass growing there in the first place. The second problem is a legal one. Household disinfectants and bleaches are not labeled as algicides and technically cannot be used as such. Copper sulfate is relatively successful at defeating algae and some brands are labeled for turf. Greenhouse disinfectants made from quaternary ammonium compounds are found to be effective, with the least amount of damage to grass plants. There is an algicide made from hydrogen peroxide and acetic acid (BioSafe Systems) that is effective and relatively innocuous in the environment but eradicating the algae is rarely a long-term solution. The conditions that allowed the algae to flourish are still there. The more permanent solution to algae problems is cultural. Creating an environment that is favorable to turf and inhospitable to algae is usually less expensive and more effective in the long term. Aerating the soil, improving drainage, balancing fertility, and providing other conditions that allows turf to flourish are more permanent remedies for algae.

The occurrence of moss on turf usually indicates a wet and cool environment, probably a low soil fertility and pH, and improper mowing. Moss is a slow-growing plant that tolerates low pH, little to no fertility, and dense shade. Low mowing is a cultural activity that encourages the existence of moss. Fortunately, moss does not tolerate wear and tear, so removal can be accomplished easily with a verticutter or other tool that will scarify the surface. Preventing moss from reoccurring involves changes to the soil that will support a healthy stand of turf. Adjustments in soil fertility and pH should be done in accordance with a soil test (see Chapter 4, Analysis). Core aeration may be necessary, as well as the introduction of quality composts to boost organic matter and biological activity. Areas subjected to dense shade may require renovation with a grass seed blend containing different types of fescues or other shade-tolerant cultivars. Trees causing excessive shade may have to be thinned. Some areas may be too shady for turf and might be better off with a shade-tolerant ground cover. Mowing turf at a reasonable height will also help to prevent an invasion of moss.

Hand-weeding may seem ludicrous on 40–60 acres of open area but there are some cases where it may be the only means of effective control. A course in San Francisco, for example, had a big problem with infestations of English daisy. Faced with a city ordinance that mandated the reduction or elimination of pesticides and the fact that no herbicide to date had been successful at eliminating English daisy, hand weeding became their only option. Although laborious (50 workers) and expensive (over $1,500 per acre), they experienced 99.9% control three years after the operation was done. The project took six days to complete (Ash, 1998). Many golfers and superintendents can overlook a few weeds in the rough and on the fairway but scrutiny on the greens is usually much more intense. At coauthor Luff's course, the cup changer spends 5–10 minutes after moving the cup (cups are moved daily during the summer months) scouting and hand-removing broadleaf weeds on the green. Dandelions, plantain, and crabgrass are

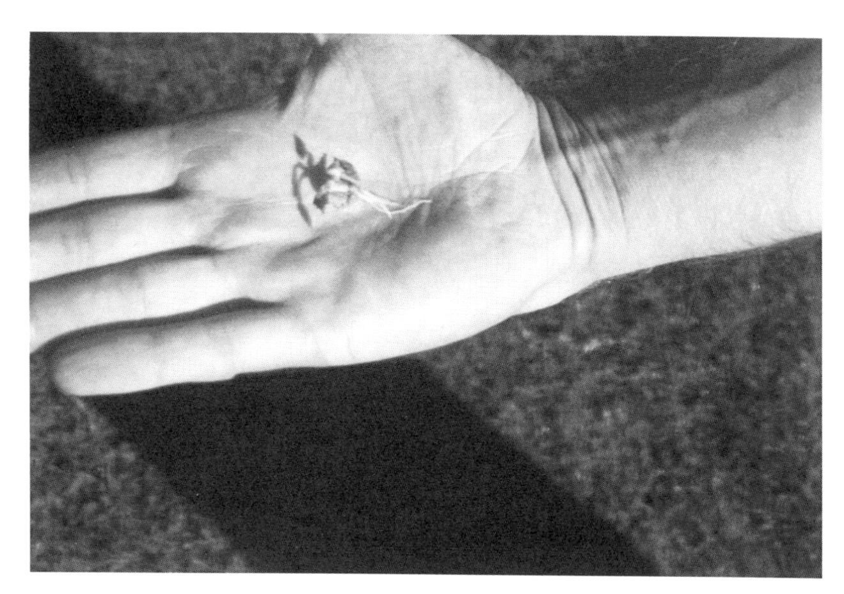

The cup changer spends five to ten minutes scouting for problems and hand weeding each green. This small investment of time has reduced herbicide use at co-author Luff's course to zero.

the weeds most often encountered. If chickweed or clover are found, the affected area often needs to be plugged. Their procedure is to first remove a plug from the nursery and insert it into the affected green. Then the plug removed from the green is transported to a section of the fairway where native bentgrasses are growing. A bentgrass plug is removed from the fairway and eventually transported back to the nursery. The plug with the weed(s) is inserted upside-down in the fairway where the native bentgrass plug was removed. Slowly, the presence of native bents increases in the nursery for future use on greens. This investment of time has eliminated the need for herbicides. Weeds on the fairways are minimal and tolerated by both golfers and the superintendent.

INSECTS

It is unfortunate that most managers have little appreciation for insects and other arthropods. Most of them that inhabit the turf ecosystem are not only beneficial, some are crucial to the mechanics of that ecosystem. Well over 90% of the arthropods that exist in the world are neutral or beneficial. That percentage is even higher in turf.

Creating a healthy and balanced ecosystem for turf would, in theory, control all pests but we are unable to accomplish this for a couple of reasons. First, we don't know how. Research has discovered a tremendous amount about the relationships that take place between the soil, plants, biomass, and atmosphere. If, however, we could fit all that we know in a thimble, what we do not understand would be hard to contain in a large truck. Secondly, even if we did know how to create the quintessential ecosystem, nature would always be throwing variables at us that we cannot control. Even a completely natural system, untouched by human management, does not respond favorably to excessive and unusual heat, drought, cold, or precipitation. Eco-complications such as these make the superintendent's job more challenging and create an atmosphere that can tempt even the most reluctant applicator to the chemicals shed. This is unfortunate, because a healthy, ecologically balanced system does provide checks and balances that are constantly protecting turf plants. This natural aegis is invisible to all but the very trained eye so its own need for protection is often overlooked.

There is ample evidence of instances where a healthy soil ecosystem initiates a chain reaction of biological, physical, and chemical phenomena that effectively controls many damaging insects—or at least limits the amount of damage they can cause. We don't know all the relationships organisms in the turf ecosystem have with plants or each other but we do know that they need certain resources to live and function. We also know that there are insects and other arthropods that offer protection to plants and that these organisms are often impacted by biocides that are applied to control target insects. It's questionable whether we will ever discover how to effectively exploit beneficial organisms to a point where insecticides are obsolete but most biologists agree that the use of pesticides can often be counterproductive.

Factors that control the success of herbivorous arthropods and their predators include climate temperature, precipitation, soil structure, biomass, and many other conditions. But the plant itself is also a consideration. There is evidence that all living things emit radiation in specific wavelengths that can be recognized by other organisms. Most insects rely on these signals for both sustenance and reproduction. According to some researchers, the insect's antennae receive the infrared frequencies given off by plants, predators, and insects of the opposite sex. Regardless of wind direction, the insect is able to locate the source of the radiation. It is thought that plants growing in suboptimum conditions generate altered wavelengths that attract grazing insects. This theory has been difficult to substantiate, but evidence of radiated signals from plants and reception of similar signals from insects does exist (Callahan, 1975). If insects begin to cause damage because of signals from subclinically ill plants, and insecticides that further stress biological functions are applied, then the turf may be rendered even more vulnerable to further attack. Treatment of the insect, which is a symptom, without analysis of the conditions in which the plant is growing, can be counterproductive and lead to further biocidal treatments in the near future.

As mentioned earlier in this chapter, plants have strong defense systems and stimulating these defenses is the least exploited approach to insect control. Mechanisms such as the production of defensive compounds, translocation of nutrients, and strengthening of cell walls depend on the condition of the plant, which is largely reliant upon a biologically active ecosystem. Fertility alone does not necessarily result in a healthy plant. In fact, excessive fertility can actually suppress plant defense mechanisms. Additionally, herbivorous insects need nitrogen more than plants. Overfertilized plants (with N) often contain higher levels of free amino acids that are known to stimulate both feeding and egg laying. Underfertilized plants, on the other hand, have a diminished capacity to defend themselves and cannot recuperate as quickly from insect damage. A biologically active soil has the ability to regulate fertility so plants are neither under- nor overfed and healthy turf is the first and most important defense against all pests.

Nitrogen is not the only important nutrient for plant defense mechanisms. Phosphorus, potassium, calcium, magnesium, sulfur, and many trace elements are also considered essential. All these elements combined, however, comprise only 5% of the turf plant's diet. Carbon, hydrogen, and oxygen from carbon dioxide and water are needed in much larger quantities than anything the superintendent can apply. Deficiencies of any one essential element can limit a plant's ability to defend itself against pestilence. Excesses can also create problems. Carbon, hydrogen, and oxygen are elements we cannot apply (with the possible exception of hydrogen and oxygen in water, which can be applied by irrigation) and it is commonly assumed that nature provides the ideal amount. Carbon, however, is not always available in ideal amounts, especially on a golf course. Carbon dioxide is diluted in the atmosphere, which is about 100 miles thick, but a turf plant growing on a green has access to only 0.125 to 0.250 inch of that gaseous envelope. Carbon dioxide produced by soil organisms seeps to the surface where the leaves of plants can capture it. When the wind begins to blow, carbon dioxide from the soil is diluted with the rest of the atmosphere. Typically, the wind is calm in the morning and early evening. The pores on the underside of plant leaves (called stomates) are, coincidentally, open during the same periods of the day. If, however, little carbon dioxide is being generated from biological activity, small plants like closely mowed bentgrass must exist on an austere diet that can limit its defense capabilities. Sand greens often cannot support large enough populations of soil organisms and their respiration of carbon dioxide is therefore limited. Without adequate carbon dioxide, the plant's ability to process soluble forms of nitrogen into protein—a form of N that is more difficult for insects to assimilate—is retarded.

Nitrogen, phosphorus, and potassium are commonly applied during fertilization. Calcium and magnesium are delivered if lime is applied. Superintendents are using iron more often to enhance turf color without promoting growth. Most other trace elements such as manganese, boron, copper, cobalt, zinc, and molybdenum are often ignored unless a leaf tissue analysis determines there is need. There are many other elements that are not considered essential to plants but are used by soil organisms. If the theory that a healthy ecosystem produces healthy, resilient plants is correct, then the needs of both plants and soil organisms should be considered (see Chapter 2, Fertility). The severity of insect damage is often related to the plant health and vigor. Stressed plants are rarely as resilient.

When insect problems become apparent, the general reaction is to identify the pest and find the appropriate insecticide to eradicate it. Applications of pesticides, unfortunately, often eradicate more than just the pest. In many cases, the insecticide used will also suppress predators that may be capable of controlling the target pests. This suppresses some of the plant's natural resistance to the pest if a resurgence of the same pest occurs. Other predators, unrelated to the target pest, may also be impacted and their inactivity may create a new insect problem. The popular insecticides used to control chinch bugs (*Blissus leucopterus*), for example, will also kill big-eyed bugs (*Geocoris bullatus*), a species of insect that is very capable of controlling chinch bugs. If the chinch bugs return, there may not be any big-eyed bugs to control them.

Dr. Daniel Potter and Stephen Cockfield from the University of Kentucky discovered another example of predator suppression (Cockfield and Potter, 1984). They found that predation of sod webworm eggs was significantly suppressed for weeks after a single application of chlorpyrifos (Dursban). These researchers put 500 sod webworm eggs out at 1, 3, and 5-week intervals after one treatment with chlorpyrifos. The number of eggs that were either eaten or carried off by predators was measured in replicated tests, and the averages are shown in Figure 5.4. For more than three weeks predator activity was significantly suppressed by a single application of the insecticide. Known predators of sod webworm eggs are ants, rove beetles, predatory mites, and ground beetles. Insecticides known to affect one or more of these insect groups include chlorpyrifos (Dursban), isofenphos (Oftanol), trichlorfon (Proxol), and bendiocarb (Turcam). Beneficial organisms such as collembola, enchytraeid worms, and saprophytic mites that play a valuable

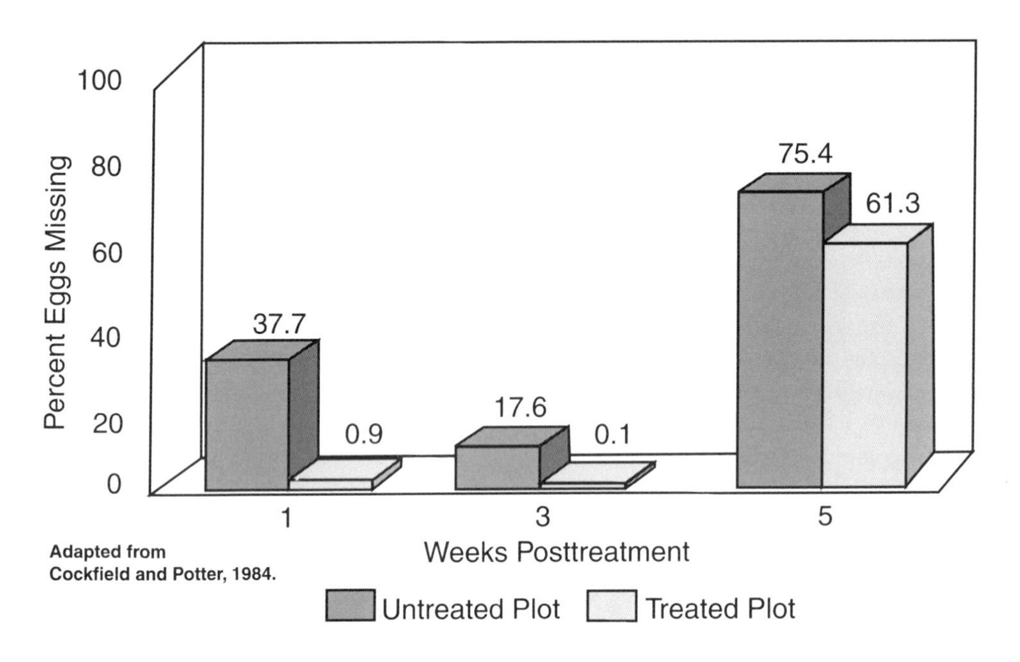

role in recycling organic matter and producing carbon dioxide—a vital nutrient for plants—are also affected by one or more of these materials.

Another test done by Cockfield and Potter showed suppression of spider and rove beetle populations (both of which are voracious predators) in areas treated with chlorpyrifos (see Figure 5.5). Both spiders and rove beetles are common turfgrass inhabitants.

The establishment of natural areas on the golf course can add beauty; reduce maintenance inputs and water consumption; and attract important insect predators, especially if wildflowers are planted. Unfortunately, there is no sure way to attract predators specific to the pest problem at hand, at least not yet. Many experiments have been conducted using flowers, pheromone lures, or sugar or nectar solutions. More often than not, predators were successfully attracted but predation was inconsistent. In one experiment using a sugar solution, large numbers of Tiphia wasps (parasites of white grubs) were successfully lured into the desired area but they preferred the lure over the prey and less predation occurred compared with an area where no attractants were used. Most researchers will agree that the more effective attractant for predators is prey and that the reduction or elimination of insecticides that cause predator mortality is a surer way of increasing their activity. Natural areas can be mowed once or twice per year to avoid the establishment of unwanted trees or shrubs.

Many inhabitants of turf and surrounding areas are insect predators or parasites. The list includes ants, spiders, rove beetles, big-eyed bugs, ground beetles, tiger beetles, green lacewings, lady beetles, predatory mites, syrphid flies, soldier beetles, fireflies, blister beetles, stink bugs, assassin bugs, minute pirate bugs, damsel bugs, ambush bugs, robber

FIGURE 5.5. IMPACT OF CHLORPYRIFOS APPLICATION ON SPIDER AND ROVE BEETLE POPULATIONS

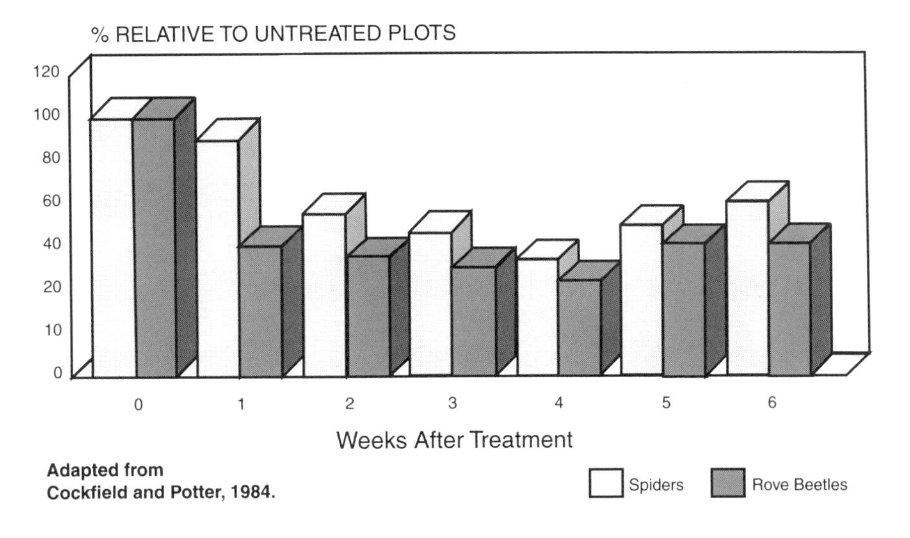

Natural areas as described in Chapters five and six. These areas are not seeded or planted. The plants are natural indigenous species that thrive on neglect.

Ants are beneficial predators of many turf pests but can sometimes cause damage. Plugging damaged areas is usually a more practical method of managing them than using pesticides.

flies, praying mantises, roaches, crickets, dragon flies, and dozens of parasitic insects such as Scoliid wasps, Typhiid wasps, and Trichogrammidae wasps. The population of these different organisms depends on resources, climate, habitat, and other conditions that may be beyond the superintendent's control. Pesticides that suppress these predators, however, are always applied by choice. Ants can sometimes cause visible damage, especially on greens, but eradication with an insecticide is usually counterproductive. When the size or abundance of damaged areas have reached beyond the tolerable threshold, they can be easily repaired by plugging with a turf repair tool. Small piles of excavated sand or soil can usually be smoothed with a drag cable or whipping pole (see Chapter 6, Cultural Practices).

There has been some research conducted that shows the impact of insecticides on different groups of soil organisms. The vast majority of these organisms are beneficial to turf in one way or another and when their activity is suppressed, fewer of these benefits are available. Slowing or stopping the flow of these biological gifts at a time when the plant is already stressed may inevitably make it even more susceptible to insect damage. As more research is conducted on the ecological side effects of pesticides, it is becoming clear that a more holistic thought process is necessary to produce and maintain healthy turf in the future.

Insects and other arthropodan pests that damage turf generally fall into one of four main categories: root feeders, stem burrowers, juice suckers, and leaf eaters. Root feeders include—among others—white grubs, ground pearls, nematodes, and mole crickets. Insects such as billbugs and annual bluegrass weevils burrow into stems and often damage the crowns. Chinch bugs, greenbugs, mites, mealybugs, and spittlebugs are important juice suckers and the leaf eaters include armyworms, cutworms, sod webworms, and other turf-eating caterpillars. Almost all of these insects are most damaging to turf during the immature stages (larvae or nymph) of their life cycle.

There are botanical or biological products that offer some control and they are relatively innocuous to the environment; however, timing is very important (as it is with conventional pesticides). Proper identification of the pest and familiarity with its life cycle are key to effective control. Almost all of the above-mentioned arthropodan pests have natural enemies that constantly offer free help to the superintendent. Care should be taken to protect this resource. These beneficials include arthropods, nematodes, protozoa, fungi, and bacteria. Pesticides that affect these groups should be avoided unless absolutely necessary and, if they must be used, it should be sparingly and only where they are needed.

Cultural practices can often result in a significant reduction of damage (see Chapter 6, Cultural Practices). Black cutworm moths, for example, lay their eggs on the leaf tips. If the superintendent's timing is right, mowing the greens, tees, and fairways with the baskets on can reduce black cutworm population by 80–90%. The key is knowing just when the female moths oviposition. Pheromone traps are available that attract males and generally, when males are being caught on a sustained basis, egg laying has begun. Black cutworms are capable of 2–6 generations per year—depending on the length of the season—so if black cutworms are a prevalent problem, vigilance is a priority.

Often, turf can tolerate damaging herds of foraging insects by engaging a defense mechanism, by outgrowing the damage, or both. Unfortunately, during the hot, dry part of the year, turf is often stressed and is unable to defend against or outgrow even a small amount of insect damage. A researcher at Michigan State University found (entirely by accident) that the amount of damage caused by many foliar feeding insects was reduced or eliminated by low volume, high frequency irrigation. Their treatment consisted of daily applications of one-tenth of an inch of water in the afternoon, during the hottest part of the day. The result was a reduction of insect damage to below tolerable levels. Coauthor Luff's crew was able to eliminate symptoms of a chafer problem with nightly light watering. If irrigation is available, using it to reduce stress makes more sense than applying pesticides, which may stress turf plants even more.

Stress also has a tendency to change the chemistry of a plant, producing a buildup of free amino acids, sugars, and other components of protein and carbohydrates that, in a free state, can stimulate insects to both feed and lay eggs. This phenomenon also occurs from nutrient imbalances, especially excess or deficient nitrogen. There is very strong evidence that a biologically active soil mediates the amount of N and other nutrients available to plants, resulting in conditions that are more ideal for turf than insect pests. Stress management can include the use of seaweed or kelp extracts that naturally contain plant hormones. These hormones not only help plants resist stress but there are several reports that clearly correlate the application of seaweed extracts with a measurable increase in the production of defense compounds. Experiments have shown average feeding reductions of 60% on plants treated with seaweed extracts compared with control plants.

Excessive thatch can provide habitat for many herbivorous insects including chinch bugs, billbugs, and many species of damaging caterpillars, cutworms, and webworms. At the same time, excessive thatch weakens turf, making it less able to tolerate insect damage. Thatch is caused when the residues of stems, crowns, stolons, tillers, leaves, and other plant parts intermingled with living roots and shoots accumulate faster than they can be decomposed. Excessive thatch can inhibit water infiltration and promote the volatilization of nitrogen from fertilizers, especially when urea is used. Thatch is often a result of poor management practices (see Chapter 6, Cultural Practices). Excessive applications of nitrogen that promote fast growth and the use of pesticides that inhibit the activities of decomposing organisms (especially earthworms) are the leading causes of thatch accumulation.

Earthworms are probably the most valuable soil organisms for thatch control. They quickly decompose surface residues and enrich the soil with castings while providing aeration, improving water infiltration, and mitigating soil compaction. Grass clippings contain an excellent slow release nitrogen that favors the existence of decomposition organisms. Excessive soluble nitrogen can sometimes acidify soil, inhibiting the activities of earthworms and other thatch decomposing organisms. Contrary to what many believe, collecting clippings can actually favor thatch accumulation. Grass clippings can be returned to greens if mowing occurs when the grass is dry and earthworm castings can easily be smoothed with a drag cable or whipping pole (see Chapter 6, Cultural Practices).

Pesticide use may play a role in the creation of thatch. Pesticides that inhibit the activities of decay organisms inadvertently advance thatch. A single application of products containing bendiocarb, benomyl, carbaryl, fonofos, or ethoprop can kill 60–99% of an earthworm population. Other compounds such as chlorpyrifos, diazinon, isazophos, isofenphos, and trichlorfon also cause significant mortality of earthworms and other saprophytic organisms. Many of these compounds also cause the reduction of certain predator groups. Pesticides used to combat disease and/or insect problems may, in the long term, only serve to increase these problems if they negatively impact beneficial organisms.

Thatch problems are relatively rare in low input or organically maintained turf because most of the inputs are as beneficial to soil organisms as they are to turf. Compost and organic fertilizers stimulate the biological activity needed to control thatch. Healthy soils generate large populations of organisms that can often consume thatch as fast as it is created. Even with organic management, however, the amount of nitrogen typically applied to greens often produces more thatch than decay organisms can consume but the rate of accumulation is significantly slower. The successful management of superfluous thatch can mitigate or eliminate other problems related to the stress it causes.

Familiarity with the habits of insect pests can often reduce their opportunities to succeed. Failure to completely fill core holes after spring aeration, for example, can sometimes result in sod webworm problems. Adult female moths find partially filled core holes a convenient place to lay eggs. Damage from larvae usually begins along the edges of the core holes. Incompletely filled core holes in the fall do not present the same opportunities. Sod webworms can usually be eradicated without insecticides by flooding the affected area with water. Some managers will mix in a little mild detergent. Within a few minutes, the webworms come to the surface where they can be dispatched with a whipping pole and, if the sun is shining, left to desiccate. Birds will often clean up the litter. If detergent is used, the turf should be rinsed to avoid scorching.

If reseeding becomes necessary because of damage from arthropods other than root feeders, a more permanent solution to many foliar feeding insects may be the use of endophytically enhanced grass seed. Endophytes are fungal organisms that live symbiotically within the cells of the grass plants and reproduce during cell division. They create a bitter tasting toxin that repels most insects and kills many of those that continue to feed. There are many different varieties of cool-season grass seeds that contain endophytes and the level of infection varies widely. For turf that is effectively resistant to foliar feeding insects, endophyte infection should be at least 70% or higher. Common sense dictates that the level of uninfected seed is relative to the potential amount of damage that can occur. If the seed has only 50% infection, then it is possible to lose half of the turf to insect damage.

Varieties of seeds that are bred with endophytes include perennial ryegrass and many different types of fescues. Natural varieties of bluegrasses or bentgrasses that contain endophytes have not been found; however, plant scientists have successfully spliced endophytes into both bluegrass and bentgrass strains. But, the strain of endophytes being used in these experiments is causing sterility, and seed production is impossible. As of this writing, endophytically enhanced bluegrass or bentgrass seeds are not commercially available. The use of endophytically enhanced fescues and perennial ryegrass may only be appropriate in the rough and, in some cases, the fairway. They may be impractical in some areas where foliar feeding insects are a problem.

If endophytically enhanced seed is stored for an extended period of time, the endophytes may die. The length of time the seed can be stored depends on the temperature of the storage facility. A refrigerated facility can keep the endophytes viable for many years but a standard, uninsulated warehouse situated in New England, for example, would only preserve the endophytes for approximately 12 months. Heat is the endophyte's biggest enemy. The longer they are subjected to temperatures above 50° F (10° C), the fewer the number of endophytes that will remain viable. Additionally, the higher the temperature, the more quickly endophyte infection diminishes. Once planted and established, endophytes will reproduce with the plant and remain active for as long as the plant lives.

Endophytes exist in the aboveground portion of the plant and do not normally provide protection from root feeding insects such as white grubs. Researchers at the University of Rhode Island, however, found that freshly germinated seedlings of endophytic varieties are resistant to Japanese beetle grubs. This resistance is greatest in tall fescues but also exists in endophytically enhanced perennial ryegrass and other fescues. The size of the grub was also a factor in these findings. Resistance was greatest against small grubs and was insignificant against mature grubs. Another university is attempting to isolate the chemical basis of this resistance, which may eventually result in a biological or botanical grub control product. The Rhode Island study also found that *"grubs are generally incapable of damaging tall fescue turfgrass in this [Rhode Island] area."* This finding has been apparent to many superintendents for some time; unfortunately, tall fescue is an inappropriate choice for many areas of most golf courses.

Grass seed infected with endophytes has shown other benefits, including improved performance. Some research has suggested that endophytes produce substances similar to plant hormones. Introducing plant hormones can also be accomplished by applying seaweed extract.

Another biological control that is sometimes effective is a commercially prepared fungi capable of attacking and destroying white grubs. Products containing the fungus *Beauveria bassiana* are labeled for turf and claim to control white grubs and some other turf insect pests. Under the right conditions, they have the potential to be as effective as chemicals. As with any biological control, environmental conditions play a major role in their efficacy.

The botanical insecticide neem has also been tested as a control for grubs. Neem is actually an insect growth regulator that interrupts the insect's maturation process, usually during molting. Unfortunately, the product has been largely ineffective on turf grubs because it cannot penetrate the soil down to where the larvae reside. Eventually, producers of the product hope to devise a vehicle that will carry the product to greater soil depths. Neem has been tested with success on other turf pests such as chinch bugs, billbugs, sod webworms, and armyworms. Products containing neem compounds, labeled for use against these insects, are commercially available.

A promising new strategy for combating grub damage and damage from many other insect groups is to apply a natural repellant to the turf before the adult insects lay their eggs. Preventing adults from ovipositing in the turf will eliminate grub damage without further treatment. Repellants containing garlic juice, capsaicin (extracts from hot peppers), or other materials that persuade insects to go elsewhere are becoming increasingly more popular with turf managers. A manufacturer we interviewed (Garlic Research Labs) recommends 13–26 oz of their 100% garlic juice concentrate per acre, mixed with enough water to drench the soil, applied every 10–30 days depending on soil type and the amount of precipitation. Sandier soils, especially after heavy rains, will need reapplications sooner. For this strategy to be effective, knowledge of the pest's life cycle and timing is crucial.

Bacillus thurengiensis (Bt) is the active ingredient in many biological insecticides that are labeled for sod webworms, armyworms, black cutworms, and other turf eating caterpillars. There are many different strains of Bt but the one most often labeled for turf pests is the kurstaki strain. Bt is a natural bacteria that is relatively specific to lepidopterious larvae because it is ingested by larvae with alkaline digestive fluids. When it is applied to the turf canopy, only grazing organisms will ingest the Bt and organisms with acidic digestive juices will not be infected. There is little danger that nontarget organisms will be affected.

Insecticidal soaps (IS) are also effective against softbodied turf pests but may be impractical for nocturnal pests. IS's mode of action is to desiccate the insect and it must make contact with the target organism to be effective—which could prove difficult at night.

Almost all white grubs are susceptible to Milky Spore disease caused by the bacteria *Bacillus popilliae* or *Bacillus lentimorbus*. There is, however, a different strain of the bacteria that is unique to each species of grub and unless the grub is infected with the right strain, it is unlikely that it will contract Milky Spore disease. Currently, and for the past four decades, products designed to infect grubs with Milky Spore disease are commercially available but they contain only the strain of bacteria that affects Japanese beetle larvae. One manufacturer we interviewed (St. Gabriel Labs) is attempting to produce almost all of the strains necessary to control almost all species of white grubs. At this point the product is not commercially available but may be in the not-too-distant future.

When a grub ingests the appropriate strain of *Bacillus popilliae*, the bacteria multiply in the larva's bloodstream and subsequently cause death. The bacteria reproduce inside the remains of the grub and increase the inoculation of the area. After two to four years, the spread of the disease is usually sufficient to keep the population of larvae below the tolerance level for turf. Depending on climate and soil conditions, this inoculation can last for more than 15 years. Experiments at Cornell University suggest that the best time to apply Milky Spore is mid- to late summer when the new brood of Japanese beetle larvae have hatched and have begun to feed. In the Northeast and Pacific Northwest spring soil temperatures are often not warm enough. Most grubs are cold-blooded; that is, their temperature is the same as their environment's. Cold grubs are relatively inactive and consumption of spores or anything else is minimal. Spores reproduce inside the grub but do so more slowly at lower temperatures and inoculation of an area located in cooler regions of the world may take longer. The shell of the spores breaks down when it comes in contact with the acidic gastrointestinal juices of the grub. Sometimes the spore's shell will not break down enough as it passes through the grub to cause infection. The shell, however, has been weakened by its exposure to the grub's gastric juices and will likely cause infection the next time it is ingested.

The most common cause of failure using Milky Spore disease is the mistaken identification of the grub. Most white grubs look very similar and the only way to differentiate between a Japanese beetle larva and other turf grubs is the raster pattern located on the tail end of their bodies. These patterns have subtle differences that identify the species of adult the larva will eventually become. Most turfgrass extension agents can provide illustrations of these patterns or they can be found in books such as *Destructive Turfgrass Insects* (Potter, 1998).

Although anecdotal, more than one turf manager we interviewed told us that when they adjusted the soil's calcium to magnesium ratio (Ca:Mg) to ~8:1, grub problems seemed to disappear. A researcher at the University of Massachusetts attempted to replicate these findings but her results were negative (Vittum, 1984). This research, however, did not focus on Ca:Mg ratio but rather on the influence of lime on Japanese beetle grub populations. Research specific to the benefits of an ideal Ca:Mg ratio is scant.

Another often effective treatment for grubs is the application of entomopathogenic nematodes. These near-microscopic worms enter the bodies of grubs and release bacteria that quickly infect and kill the grub. Nematodes usually enter the insect larvae through natural openings and the bacteria they release produces a toxic enzyme which kills the grub within 24 hours. The nematodes then use the remains of the grub as resources for their breeding activities. Usually, those resources will support the conception of three successive generations of nematodes, after which the nematodes must escape the body of the host and find new larvae to infect. At soil temperatures around 70° F (20° C), this cycle takes about 1–2 weeks. The infective stage of the nematode is called the juvenile stage. Juveniles do not eat and can live in the soil for extended periods of time.

Nematodes need moisture to remain active. If they are introduced into a dry soil, or if the soil dries out after they are applied, they are likely to be ineffective at controlling grubs. If they are applied in the spring and the soil dries out during the summer, they may have to be reapplied in the fall if they are to control the newest brood of larvae. It is best to thoroughly moisten the soil before they are applied. Nematodes can migrate into the soil if there's adequate moisture. There is evidence that nematodes will accumulate after several annual applications and continue to control grubs without reapplication but it is not known for how long.

Nematodes function best in loamy soils where there is plenty of porosity but not so much that the soil dries out too quickly. Heavy soils inhibit the movement of nematodes, but grubs are rarely found in these soils so it is somewhat of a moot point. Sandy soils, on the other hand, can harbor large infestations of grubs but often do not stay moist enough to sustain active nematodes. Unless there is irrigation or the climate is naturally moist, nematodes may not perform well in a very sandy soil. If there is adequate moisture, most juveniles can exist for about three months in the soil. There is evidence that nematodes can enter a dormant stage to survive conditions such as drought or frost. Introducing them as protection from grubs before grub activity is present is a common practice. Most nematodes can exist in a temperature range of 32 to 90° F. They are, however, most active in soils that are between 65 and 85°. Turf that is mowed too low during the hottest part of the season may allow soil temperatures to exceed the upper limits of the nematodes' temperature range. Some strains of nematodes can migrate down into cooler horizons, some cannot.

Nematodes can withstand high pressure—up to 300 psi—which makes it possible to apply them through almost any water dispensing system. They are best applied in the early morning, early evening, or on a cloudy day. Exposure to direct sunlight in excess of seven minutes can sterilize them. They will still be able to infect and kill grubs, but they will not be able to reproduce.

There are many strains of nematodes that are entomopathogenic but only a few are commonly used in commercial offerings. The *Steinernema* sp is a strain that attaches itself to a soil particle and waits in ambush for a grub. There is some research that suggests this strain is less effective at controlling grubs because these nematodes reside too close to the soil's surface. A strain called *Steinernema glaseri* shows greater promise as an effective, biological grub control. The Steinernema nematodes reside near the surface of the soil and attach themselves to hosts that frequent the soil/thatch interface. Webworms, cutworms, armyworms and wood-borers are particularly susceptible to this strain of nematode. Steinernematids are most effective at temperatures between 72 and 82° F (21–26 ° C) The *Heterorhabditis* sp is a hunter which can seek out grubs by following their trail of exudates. Unlike the Steinernema strain, it can find and infect larvae several inches below the soil's surface. The Steinernema nematodes reproduce via a male and a female entering the grub whereas the Heterorhabditis nematodes are hermaphroditic and only one is needed inside the grub to initiate the reproductive cycle. Each of these strains can be effective under the right conditions and preparations that include both *Steinernema* sp and *Heterorhabditis* sp are available. There is a complete list of suppliers in the most recent Directory of Least Toxic Pest Control Products, produced by BIRC, P.O. Box 7414, Berkeley, CA 94707, phone 510/524/2567, E-mail: birc@igc.org. This directory contains sources of many different biological and botanical pest controls.

Superintendents who have problems with mole crickets will also be able to use entomopathogenic nematodes soon. A new strain, recently discovered in South America—where mole crickets originated—is capable of effectively controlling them on a large scale (Roth 2001). Researchers at the University of Florida, University of Nebraska, and Ohio State University have been experimenting with several different strains including *Steinernema riobrava, S. scapterisci*, and *Heterorhabditis indica* and found that *S. scapterisci* provides the most effective control against mole crickets. The University of Florida has secured three patents for using them as an insect control and they expect the nematodes will be commercially available soon.

Entomopathogenic nematodes are different from pest nematodes (phytopathogenic) that attack plant roots and they cannot change from insect pathogens to plant pathogens. There are, however, some species of nematodes that do not normally feed on roots but will if their preferred source of food is not available. Tylenchus nematodes, for example, normally feed on fungi that may or may not be pathogenic to plants but the nematode can change from being a beneficial nematode to a pest if soil fungi are not available. In a biologically active soil, Tylenchus would never be a significant pest but might provide many benefits to plants. The Tylenchus might mistakenly consume the Chytrid fungi that becomes a parasite when ingested by a nematode. Many fungi, in fact, release the same materials as plant roots to attract and trap phytopathogenic nematodes. High populations of organisms living in the rhizosphere, such as mycorrhizae and bacteria, normally offer adequate protection from pest nematodes. The Mononchus nematode feeds primarily on other nematodes, many of which are pests. In a biologically rich soil, pest nematodes should never be a problem. In view of the fact that the vast majority of nematode species are beneficial—some crucial—to a functioning soil ecosystem, using a nematicide is almost always counterproductive.

Many turf managers often ignore the simple philosophy that a problem should be found before treatment is applied. Integrated Pest Management (IPM) has, to some, become just another commercially attractive buzzword like low spray, no spray, and natural. In a true IPM program, treatment is usually withheld until the problem and the problem area is identified. In order to do this properly, monitoring or scouting has to be part of the program. The problem with monitoring, according to many professionals, is that it is too time-consuming. Unfortunately, what many superintendents fail to understand is that monitoring or scouting will, most often, pay for itself (and often result in significant savings).

A great example of this is a project conducted by Cornell University in 1992 at a golf course in Rochester, NY. Workers scouted the entire course for grubs by lifting sections of sod with a cup cutter in a 10-foot by 10-foot grid pattern. The cup cutter removes a plug that measures one-tenth of a square foot so an estimated number of grubs per square foot could be calculated. They identified all the problem areas on the course for a labor expense of $360. The amount of money that they saved from treating only the problem areas exceeded $60,000 in the first year and the course did not sustain any damage from grubs. The scouting program was an immense success. A program like this would net higher profits whether a chemical, botanical, or biological material is used to control the grubs.

There are other good examples of how monitoring has not only reduced the number of problems on a course but has also reduced the cost of maintenance; however, the superintendent's crew needs to be trained to identify pests and tolerable thresholds. This training may subsequently lead to higher wages for staff members but the outcome is better insect control and more savings for the course.

Another Cornell University research project (Grant et al., 1994) monitored grub populations on more than 300 residential lawns in upstate New York and found that even in a bad year (good for the grubs), only 18% of the total lawn area monitored needed treatment. They also found that most preventative treatments for grubs were a waste of time and money. If, on average, a manager could cut back the amount of grub controls by 82%, the savings would be significant.

The use of light traps can be a very helpful tool to indicate what types of insect pests are visiting the course, especially at night. Early indications of various pests can prepare or alert managers to scout more frequently and apply controls before significant damage has been done.

Finding problems on the course before they become epidemic is always the most prudent practice, for both esthetic and economic reasons. Treating problems that do not exist can very often be a waste of time, effort, and money, and an unnecessary infusion of pollutants into the environment. It can also alter the ecology of the treated area to a point where new problems are created.

If grub damage is severe, renovation may be the only solution. Disturbing the soil with a tiller before seeding has its advantages and its disadvantages. The tiller can act as a physical killer to grubs, literally beating them to death, but the tool can also introduce excessive oxygen into the soil which can destroy a good portion of the soil's organic matter component. In a rich soil the trade may be worth it, but in an already poor soil, the loss of fertility may jeopardize the health of the new seedlings to a point where maladies other than grubs affect the turf. Tilling in good quality, mature compost is an effective way to counter this problem. Mature compost, if incorporated into the top four to six inches of the soil, can be used at doses of one to two cubic yards per thousand square feet. Too much, however, can create a layer of soil that makes too abrupt a change in consistency from the layer beneath it. This condition (called layering) can inhibit the movement of water and gases through soil horizons. Another problem with the tiller is the amount of dormant weed seeds brought to the surface by deep tillage. Shallow tillage brings fewer seeds to the surface. Additionally, if well-made, mature compost is incorporated, the speed at which turf seeds germinate is generally increased and seed mortality usually decreases, especially when seaweed extract is also applied. The turf canopy often establishes quickly enough that weeds won't have the opportunity to become competitive. Slit overseeding is another alternative that will not disturb the soil significantly, nor cause the destruction of organic matter. This tool will physically kill some of the grubs but it doesn't cut deeply enough to cause significant impact. Core aerating will also adversely affect a grub population. Topdressing with a compost sand mix will replace the organic matter lost by oxidation.

There are many cases where the population of grubs is not concentrated enough to damage turf, but adequate to attract skunks, moles, raccoons, crows, and other animals that are foraging for larvae. The damage caused by these creatures digging through the turf can be significant. The immediate response by many turf managers is to apply an effective larvicide as soon as possible. Unfortunately, this practice is often a waste of time and money. The first problem is that these animals may be foraging for something other than grubs. The second is that by the time the larvicide has taken effect, most of the animal damage may have been done. And a third factor to consider is that when these animals are most actively foraging, the grubs are usually at their most advanced instar stage—a stage at which most larvicides offer little or no control.

These animals are great indicators of problem areas on the course. Notes can be taken or maps drawn indicating hot spots. Over the years, this information may prove extremely valuable. It is helpful to know where problems areas are if treatments or repellants are going to be used successfully in the future.

Skunks, when they are not digging for grubs, are generally beneficial. Other favorites in their diet include grasshoppers, crickets, cutworms, and other fleshy insects such as sod webworms and caterpillars. They also feed on bird eggs, moles, shrews, and small reptiles. Skunks nest in wood or brush piles, hollow logs or trees, or stacks of lumber or firewood. Removal of nesting sites such as these can reduce the amount of skunk activity on the golf course.

Similarly, raccoons like to nest in abandoned burrows, hollow trees or logs, or other covered and cozy refuges. Elimination of the nesting sites will usually result in a reduction of raccoon activity.

Moles are also beneficial animals that can cause unintentional damage to turf. It's unlikely that controlling white grubs will have much effect on mole activity because they eat other insects and earthworms. The raised ridges created by moles are unsightly and can cause turf root mortality if left unattended for too long. Moles are solitary creatures so there are usually not more than three to four in an affected area. Trapping is the most effective method of eliminating moles but this strategy may not be practical on a golf course. None of the folklorish remedies like chewing gum, car exhaust, lye, razor blades, thorns, household chemicals, gasoline, broken glass, human hair, or flooding the tunnels have any significant effect on moles. Many of these so-called remedies are much more harmful to turf than the moles are. The unsightly ridges created by moles can be flattened by the roller of a walk-behind greens mower or any other type of roller. The few minutes per day it takes to repair damage from moles is less costly, less time-consuming, and has less ecological impact than applying a larvicide for grubs or setting (and checking) traps. There are a few commercially available products containing castor oil as their active ingredient that have shown promising results. These products need to be watered into the soil to be effective. Healthy, well-rooted turf can often mask the damage caused by moles. In a thick stand of turf, moles don't usually break the surface because it is held together well by turf roots. However, the healthy, thick stand of turf may have fewer meals for moles anyway.

Treatment for animal damage can consist of a rake, a roller, possibly some seed and some well-aged compost or organic fertilizer. In a healthy turf, the evidence of grub hunters should be gone within a few days. Conventional treatment for grubs may stress the biology of the area and inhibit its ability to recovery as quickly.

Birds, such as starlings or crows, that appear as pests on the course may be performing a valuable service by foraging for insect larvae or adults. There is no guarantee that what they are eating are, in fact, turf pests, but if insect pests are abundant, a large portion of the bird's diet will likely consist largely of that insect. In addition, some birds, like starlings, can act like an ultralight, silent running, microfine aeration machine, pecking tiny holes in the sod where air and water can more easily interface with the soil. Crows can sometimes cause more damage than the four-legged diggers. They rip and tear up turf in large chunks and rarely find enough grubs to justify the amount of damage they cause. Their foraging, however, is a pretty clear signal to the superintendent that a closer look is advisable. Crows can wreak havoc on an area for as long as plump tasty grubs are close to the surface and about the only course of action is to repair the damage as often as needed. Crows tend to work in the same area but their schedule is often inconsistent. During some periods repairs will be necessary on a daily basis and then the crows will take a day or two off. Just when it seems as though they've gone elsewhere, the damage reappears. Treatments for grubs are ineffective for the most part because when crows and other animals are digging for grubs, the larva are usually too big to control. If grooming, fertility, water, and patience are applied, chances are good that the turf will recover stronger than before. If the superintendent is not willing to tolerate the damage component of this cycle, however, then the bird's target insect must be identified and treated well before the foraging begins. Action requires prior knowledge of the susceptible areas and of the insect pest's life cycle. Larvicides need to be used at early instar stages and repellants before eggs are deposited. Eradicating the birds, by whatever means, is counterproductive and, in many regions, illegal. Unfortunately, controlling the insects with pesticides may cause a disruption to the ecology of the immediate environment that may, in the long term, cause further problems. Again, the strategy of repelling the adult females before egg laying occurs may be just the ticket for eliminating both grub damage and damage from grub hunters.

The best winged hunter we've seen is the robin. It will wait patiently and listen for activity beneath the soil's surface. When the location and depth of its prey has been determined, it plunges its beak into the soil and almost always extracts some food.

If bird populations become a real problem, researchers from the USDA found that grape flavoring from a food grade chemical called methyl anthranilate is an effective repellent for many birds. Products containing this ingredient can be diluted and sprayed onto problem areas and the birds generally find it very distasteful. There are a couple of EPA registered products on the market (ReJex-iT®AG-36 and ReJex-iT® TP-40) that contain this active ingredient.

Many superintendents welcome winged predators to the golf course and some, in fact, erect bird and bat houses to attract more of these hungry hunters. Although inconsistent and difficult to quantify, predation has a dollar value and it makes economic as well as ecological sense to exploit it as much as possible.

Activities of beneficial pests such as birds, earthworms, moles, skunks, and other organisms are largely beneficial to their immediate ecosystem but may cause some esthetic impact on the landscape. Unfortunately, eliminating these pests can often result in infestations of other, far more damaging pests.

THE ECONOMICS OF ECOLOGY

There are many strong arguments for and against the use of pesticides and, generally, they all make sense to someone. People who are concerned with the environment or human health may be inclined to believe almost anything negative about pesticides. Generally, they are convinced that pesticides pollute the environment and threaten human health. On the other hand, people whose livelihoods are deeply entrenched in the purchase, sale, or application of pesticides may believe almost any positive statement ever made about these chemicals. Most of them believe that environmental and human health concerns are nonissues and that there is no hard evidence to substantiate these claims. The issue has evolved into a political tug-of-war. Standing in between these two opposing groups are people who don't understand, don't care, or both. We believe most golfers stand in this group. If the superintendent can provide them with a course that looks good and plays well, golfers probably don't care how it is maintained.

In science, especially turf science, discussions of tolerance are ubiquitous. The amount of tolerance any course has for any pest ultimately depends on the decision-makers. If acceptable levels of weeds or insect damage is low (i.e., they cannot tolerate much) then that becomes the tolerance threshold of the course. The lower the threshold, the more intense the management regime must be. The board of governors may not understand ecology but can usually understand economics. If it is apparent to them that their higher expectations carry relatively higher costs, they may be inclined to tolerate a more natural looking golf course.

The most highly and consistently used pesticides on most golf courses are fungicides, accounting for 80% or more of total pesticide use. Fungicides can cost anywhere from $200 to $400 per acre plus application costs. An average 18-hole course has two to four acres of greens, four to five acres of tees, and 30 to 40 acres of fairways. For comparison purposes, we asked the State of New Hampshire for commercial pesticide statistics on golf courses. The data clearly revealed vast differences in pesticide usage from course to course. On average, courses that could afford more intensive management consistently treated 100–200 acres of turf per season and, depending on the season, some were as high as 250 acres. Lower budget courses applied considerably less. These courses applied pesticides to an average of 30–50 acres of turf. We calculated acres by the actual coverage of all applications. In other words, if a superintendent treated one acre twice, we would count it as an application to two acres. It is easy to see how expensive pesticides can become for a golf course. A course treating 150 acres at $300 per acre is spending $45,000 on pesticides. Reducing pesticide use is not only ecological, it's economical too. Coauthor Luff's course located in the coastal New Hampshire region treats, on average, five to eight acres per year and strives for less each year. Golf courses are businesses and need to show a profit like any other. The money spent on many pesticides is unnecessary and clearly decreases profits. From an ecological perspective, pesticides may impact the balance of soil relationships to the point where the need for more control is perpetuated. The fear that all will spoil as the biocidal tap begins to close is endearing to most pesticide manufacturers but the reality is quite different.

Chapter 6:

Cultural Practices

Knowledge is a valuable asset but realizing there is more that is unknown than known is a humbling and important awakening for most professionals. Most experienced superintendents realize this. Decades of accumulated experiences and wisdom, earned from countless successes and failures register in the confident but humble expression of the veteran superintendent. Knowing the right thing to do and the right time to do it is a product of this experience, an application of skill—an art. This greenskeeper has learned to overcome the urge for reckless reaction in favor of thoughtful response. He comprehends the anomalies of his course and its ever-changing conditions. He recognizes that products and practices that work elsewhere may lack efficacy at home. He understands that research and anecdotal evidence sometimes lead to innovations but often need to be customized to fit his unique setting. If he wants to reduce or eliminate pesticide use, however, he also needs to consider the ecology of his environment and recognize how crucial it is to his decision-making process.

The information in this chapter includes ideas and innovations practiced by different superintendents and researchers who favor adjustments in cultural methods over applications of biocides. Some may be practical and successful for the reader and some may not. Some may have to be adjusted or tuned to fit the circumstances of the course where it is to be applied. Some may need to be considered and disregarded. If only one idea from this chapter reduces the need for pesticides on the reader's course, this book will have accomplished its goal.

The practices that cultivate beautiful and healthy turf are not all learned in the classroom or the textbook. Many are learned in the field, usually with the guidance of a master who's spent years honing his skills well beyond what schools and research could offer. His humility is as valuable a skill as any other he's learned. The superintendent is often weighed down with an overwhelming amount of information from multiple sources and must sift and cull until the ideas that make sense pass through his discriminating filters. Even then, there's no guarantee of success and often no surprise if he is humbled yet again. Skepticism becomes an automatic response to the latest and greatest of anything. The experienced super takes information, whether it's from a university, another superintendent, or a salesman, at face value. He understands that no matter how successful a procedure is elsewhere, it needs to be tested and probably adjusted before it has real value to him. It's the innovative superintendent who experiments with different methods that make sense but he doesn't confuse himself by testing too many hypotheses at once.

Successful cultural practices include what the superintendent doesn't do as much as what he does do. If there is intent in an omission, it is a cultural practice. Leaving the clippings on the fairway or even the green is a cultural practice if it's intentional. Not irrigating or applying fertilizer or skipping a mowing cycle during heat stress periods or disease outbreak is also a cultural practice if it is omitted on purpose.

Good ideas that work are often shared. Superintendents and researchers who've had success with different cultural practices often write about their experiences and there's usually a turf, golf, or landscape periodical somewhere that will publish it. Searching all the journals and trade magazines for interesting ideas can easily absorb too much valuable time but there is a database at Michigan State University called the Turfgrass Information Center and they have collected an impressive number of records that pertain to different articles, books, etc. The Turfgrass Information Center can be located on the web at http://www.lib.msu.edu/tgif. (For more contact information, see Sources and Resources.)

Most modern superintendents have many expensive tools in their arsenal that are designed to create better, faster playing surfaces and accomplish it in less time. Some of these specialized tools are used as often as daily, and others as infrequently as annually but all are considered essential. Not only are all of these tools necessary but the way in which they are used is critical. Every superintendent who's worth his salt knows the dangers of poorly adjusted or improperly used mowers, sprayers, spreaders, aerators, or verticutters. Almost every tool used on the course has the potential to beautify or create an unmitigated disaster. Proper use and accurate adjustments are critical but not normally difficult for someone with experience. It is important, however, that the operator of these specialized tools consider their impact on the ecosystem that promotes healthy, disease-resistant, and competitive grass. Many beneficial practices can become counterproductive if done improperly, too often, or not often enough.

There is probably nowhere on earth that golf course-like conditions exist naturally so it would be difficult to peek at how nature cultivates a healthy green or fairway. On the other hand, most living organisms—including plants—have an innate ability to adapt to environmental conditions as long as certain basic needs are met. Grass plants grow in a system that includes the plant, the soil, and the atmosphere. The dependence that each has on the other can be enhanced but cannot be replaced by man-made inputs, especially if the superintendent is trying to maintain the course ecologically.

Organic matter is home, food, and energy for billions of essential organisms in the turf ecosystem. Inadequate resources for these organisms can result in a reduction of biological activity and an increase in problems that they often suppress. Cultural practices that reduce this asset increase the likelihood of problems. Compaction, for example, is a constant problem for most superintendents. A compacted soil lacks pore space that would otherwise contain both air and water, constituents that are not only vital to healthy plants but extremely important to beneficial soil organisms. Mowing with heavy equipment—especially when the soil is wet—and collecting clippings (food for earthworms and other decay organisms) contributes to the natural inclination toward compaction. Overuse of aeration tools can also contribute to long-term problems if the introduction of too much air is oxidizing valuable organic matter and it is not being replaced. This is generally less of a concern to most superintendents because the importance of organic matter is overshadowed by the importance of having a well-drained playing surface. Superintendents who understand the value of soil organic matter in the greens' soil usually have less frequent and less severe turf problems.

The creation of soil organic matter is a very slow biological process (see Chapter 1,

Soil Ecosystem) and there are many different environmental conditions that control its accumulation and properties. Regions with warmer climates and longer growing seasons naturally have less soil organic matter than colder climates with shorter seasons. Dryer regions generally have lower natural levels of soil organic matter also. Aside from the natural conditions that influence the accumulation of soil organic matter, there are also cultural practices that can accelerate its demise. Organic matter can include everything from the living biomass to stable humus with thatch, plant residues, and friable humus in between. For the purpose of this discussion, we are not referring to superfluous thatch or plant residues when we write about soil organic matter. We are alluding to stable, friable, and other forms of humus. The first step toward preserving this very valuable asset is to understand some of the many benefits it provides (see Chapter 1, Soil Ecosystem). One of the main objectives of this book is to convince the reader that soil organic matter and the diversity of life it supports are vital components of a healthy and functioning turf ecosystem. A crop that needs as much attention as the turf is the crop of organisms living in the soil. These organisms contribute immeasurably to the health and welfare of turf plants. Their ability to manage nitrogen, for example, makes them indispensable. Applying too little or too much nitrogen is too easy for even the most experienced superintendent. The factors that influence both the plants' need and their utilization of nitrogen are constantly changing and unpredictable. Soil organisms have the innate ability to regulate the amount of nitrogen available to plants in an almost perfect synchronism with plant needs. The same is true with most other soilborne nutrients. How is it possible that soil organisms can regulate nutrients without knowing, in advance, what different plants need? Believe it or not, plant roots release nutrients that foster populations of specific organisms that are best suited to serve plant requirements.

Contrary to what many believe, adequate levels of organic matter do not compromise drainage or playability but can provide a plethora of benefits, many of which reduce stress or increase the turf's resistance to stress.

Some of the factors that influence the accumulation and characteristics of organic matter include oxygen, water, temperature, pH, and the carbon to nitrogen ratio of the soil. These factors also influence plant growth which is the main source of residues contributing to the formation of soil organic matter. In general, conditions that are optimal for the plants are also best for the accumulation of organic matter. However, optimal means *most favorable, desirable, or satisfactory*. Excessive levels of anything good for plants can cause problems. Most of us are well aware of problems caused by too much water or fertilizer but many other materials and cultural practices done to excess can also cause problems.

Indiscriminate applications of lime, for example, can raise the soil pH to levels that are more favorable for decay organisms but there may not be a corresponding increase in the production of plant residues (i.e., plant growth is not increased). The resulting increase in their population can cause a related decrease in soil organic matter. Superintendents who lime without the benefit of a soil pH test run the risk of creating this scenario. In Figure 6.1 it is evident that elevated levels of lime increase the evolution of carbon dioxide from the decomposition of soil organic matter. A soil pH test should be the only criterion for applying lime (see Chapter 4, Analysis). Constant and prolonged use of calcareous sand can also raise the soil's pH to above optimum and correction may be difficult if not impractical. Regular contributions of well-made, well-aged compost during topdress operations can replace organic matter losses from a higher than optimal pH and buffer changes in pH from calcareous sand.

FIGURE 6.1. INFLUENCE OF LIME ON CO_2 EVOLUTION FROM DECOMPOSING ORGANIC MATTER

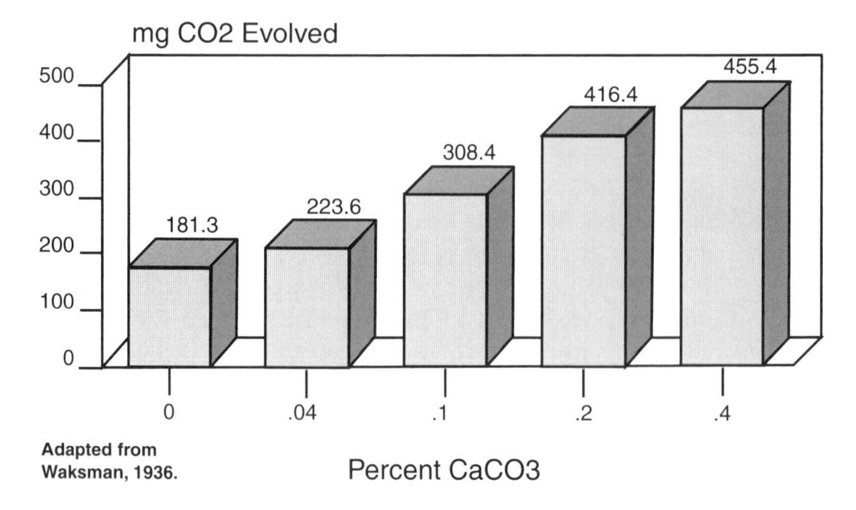

With so much attention paid to drainage, many golf courses start out with a conspicuous absence of organic matter in the soil and superintendents seem to keep it that way, whether intentionally or inadvertently. The natural porosity of sand increases the volume of air that resides in the soil, reducing the likelihood of humus accumulation. Core aeration, which is often essential to alleviate compaction, increases oxygenation even more. Usually more sand is applied during topdressing operations. Many superintendents use a mixture of sand and a small percentage of topsoil as a topdressing material. Unfortunately, sand contributes no organic matter to the soil and the quality of the topsoil used in these mixes is rarely tested for organic matter or anything else.

Imported topsoil is usually altered in the process between excavation and installation. The digging, scraping, dozing, loading, transporting, dumping, mixing, screening, and spreading of topsoil causes significant changes to its structure, chemistry, and biology. If topsoil sits in a pile for any extended period of time, more of these changes will occur. Unfortunately, none of these changes improves the quality of the topsoil. Organic matter content is reduced, beneficial organisms such as earthworms and mycorrhizae fungi are all but wiped out, and the aggregation of soil particles is significantly diminished. The biological component will recover and eventually repair much of the damage if the topsoil contains optimum levels of organic resources but the organic matter component is significantly reduced between the time it is excavated and when it is delivered. The only component of topsoil that may remain unaltered is the weed seed content.

A well-made and well-aged compost substituted for topsoil in the topdressing mix can not only arrest the loss of soil organic matter, but can provide scores of other benefits as well, including disease suppression, thatch reduction, stress resistance, and reduced water consumption without compromising drainage. A topdressing mix with compost is especially helpful after core aeration; filling the core holes with a compost mix can make soil less prone to compaction. Mixtures as rich as 1:1 (sand:compost) can gradually introduce enough organic matter to resist compaction on a more sustained basis.

A recent study from Rutgers (Rossi, 2000c) suggests that core cultivation gives only

The topdressing phase is almost complete. The topdress material is ready to be brushed into the turf.

temporary relief from compaction; that within three weeks of aeration, compaction often returns to pretreatment levels. Returning clippings to greens that are prone to compact encourages populations of earthworms that can alleviate compaction on a daily basis. Their tunneling activities create passageways through which air and water can infiltrate. The castings left on the surface of the greens can be easily smoothed each morning with a drag cable or whipping pole. Unfortunately, clippings can only be left on the green if they are dry. Wet clippings clump and create insufferable challenges for golfers. If problematic greens can be mowed late in the afternoon or early in the evening, the dry clippings left behind will rarely be noticed by golfers. It is important, however, to consider the stress factor when mowing in the afternoon. During the hot summer months, earthworms are relatively inactive and mowing during the driest part of the day can also be the hottest part of the day. The stress of mowing added to heat- and perhaps drought-stress might be the proverbial straw that breaks the camel's back. If practical, mowing in the early evening may be the most ideal time for recycling clippings on stressed greens.

Other benefits from recycled clippings include contributions of nitrogen, better water infiltration, improved color, disease suppression, crabgrass suppression, increased root production, and thatch reduction. Over half the nitrogen from the fertilizer we apply resides in the clippings that we often remove and discard. This return of organic nitrogen stimulates biological activity that cycles valuable carbon dioxide back to plant leaves for carbohydrate production, which increases root growth. Color improvement and the suppression of disease and crabgrass are secondary side effects. Over 25% more root mass is often produced where clippings are returned which, on a dry basis, can contribute one-half to four tons of organic matter annually per acre. As the moisture evaporates from fresh clippings, the surrounding area is cooled, mitigating heat stress. If recycling clippings on greens is impractical under any circumstances, their value for composting should not be ignored. All the benefits mentioned above are retained and enhanced in the compost process and can be applied when topdressing.

Brushes work topdress material into the turf canopy. This operation is usually completed in 2-4 passes, each in a different direction. Brief irrigation after brushing moves any leftover material through the canopy and mitigates stress.

Height of cut (HOC) also has a profound effect on the production and accumulation of organic matter. Root mass is directly affected by the height a superintendent chooses to cut his greens. Roots grow from the sustenance they get from photosynthates produced in the leaves. The greater the leaf surface area exposed to the sun, the more photosynthesis can occur and the more nutrients become available to roots. Figure 6.2 shows that for each eighth of an inch the mower is raised, there is a 30% increase in the amount of leaf surface area exposed to the sun. Deeper and more divergent root systems allow better, more efficient transport of water and nutrients through the plant. Improvements here also increase resistance to drought, increase internal cooling in hot weather, and reduce root loss that is typical during summer months. The resulting mitigation of

FIGURE 6.2. INFLUENCE OF MOWING HEIGHT ON TURFGRASSES PHOTOSYNTHESIS

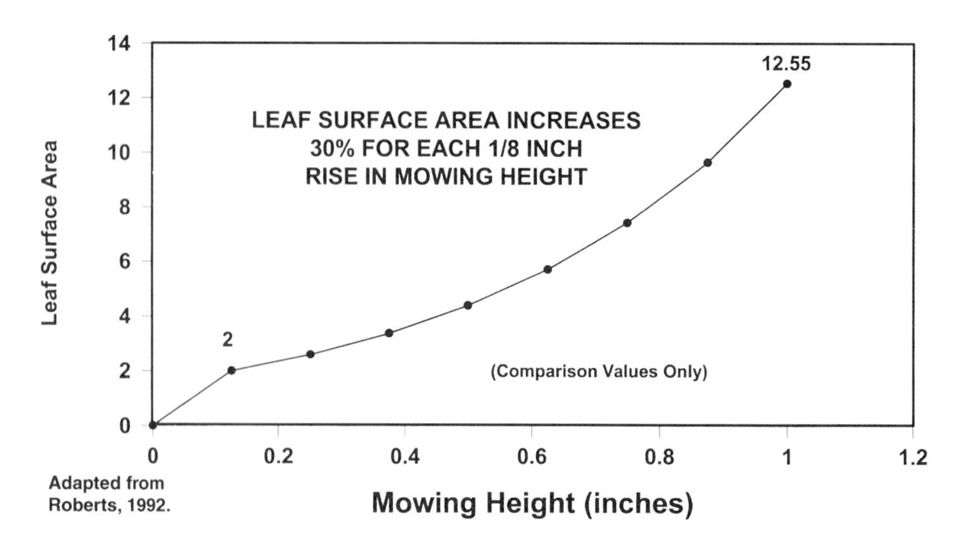

Adapted from Roberts, 1992.

Attaching a steel brush in front of a walk-behind greens mower also improves the cut at a higher HOC.

stress decreases the plant's susceptibility to opportunistic pests and maintains the production of valuable organic residues at optimum levels.

The height at which greens are cut also affects ball speed. If it didn't, there would be few superintendents who would even consider mowing their greens at one-eighth (0.125) inch (or lower). Unfortunately, there is usually a direct relationship between green speed and the frequency and severity of problems. As the HOC is decreased, problems seem to increase. If player pressure has already reduced the HOC to below an ecologically reasonable level, it's unlikely the mower will be able to be raised much without objection. Rolling, however, can help maintain green speed at a slightly higher cut. Rolling periodically instead of mowing during certain times of the season can help turf maintain vigor, but rolling—or any other mechanical cultivation—is not recommended when greens are wet or severely stressed. Verticutting during heavy growth periods thins the turf, reduces grain, and increases ball speed. Light, frequent topdressing and double cutting will also increase ball speed. Verticutting or double cutting are not recommended during stressful periods of the season. The need for well-aged compost on high speed, high performance greens is greater because there is little organic matter (with the exception of thatch) being produced by these extremely stressed plants (see Chapter 3, Compost).

Attaching a steel brush in front of a walk-behind greens mower also improves the cut at a higher HOC. The brush should be mounted on a hinged frame so that it can be disengaged easily. The brush will have the opposite of the desired effect on wet grass—laying down the shoots instead of standing them up. Brushes should be suspended so that they do not exert too much abrasive force on the turf canopy and should not be used when grass is severely stressed.

New varieties of bent grass bred to thrive at extremely low HOCs are attracting a lot of interest. These new cultivars—called *high shoot density cultivars* (HSDC)—grow at extremely high densities (300–400 shoots per inch^2) and seem to tolerate normal wear at cutting heights at or below one-eighth (0.125) inch but are not without problems. The highlights of HSDC are the upright growth characteristics and their capability of maintaining shoot density throughout the season. During the first year of growth, HSDC develop a greater and deeper root mass than conventional bents but in subsequent years there is no significant difference. Early reports of increased disease susceptibility in certain areas and excessive thatch accumulation have surfaced but HSDC have only been available for a short period of time and research continues. Cornell University test plots haven't had a problem with thatch but they religiously topdress every two weeks with medium-fine sand. The slow, dwarf-like growth characteristic of these new varieties impedes recovery from various injuries such as divots and ball marks. Compared with conventional cultivars mowed at the same height, there appears to be no significant difference in ball speed. Establishing HSDC in existing turf is proving to be no easier than making the seamless and invisible transition into any other bentgrass. Experiments at Ohio State University showed that the six different varieties of HSDC they were testing were unable to establish themselves at all, even accompanied with a slew of cultural practices designed to suppress the existing turf. Apparently HSDC do not compete for sunlight very well against conventional bents. The researchers concluded that if a superintendent wants these new cultivars on his greens, his best bet is to start from scratch. At first glance, it doesn't look like these new varieties will require any less management. Indeed, most researchers are finding that HSDC need more intensive management including more topdressing, more verticutting, and more double cutting. Intensification of these practices, especially during stressful periods of the season may lead to other unwelcome headaches for the superintendent.

Soil temperature can have a profound effect on the fate of soil organic matter. As illustrated in Figure 1.5, at a soil temperature of 88° F (~31°), with adequate air and moisture, soil organic matter is destroyed faster than it can accumulate. This is a common condition in tropical and subtropical soils, where high temperature, moisture from tropical rains, and the abundance of air in extremely porous soil are all at an optimum level for decay activity. There is a direct correlation between the average annual temperature of a given region and the native levels of soil organic matter. As one moves closer to the equator, it is evident that the natural existence of soil organic matter lessens (see Figure 1.4). Even in a tropical rain forest where prolific plant growth produces an abundance of organic residues, very little organic matter can accumulate in the soil. At lower cutting heights, the soil has less protection from the sun's heat. Elevated soil temperatures not only stimulate more biological activity but also stress plants and reduce growth that inevitably contributes to the creation of soil organic matter.

Shade can mitigate the effects of high temperatures on soil organic matter but shading turf can be counterproductive. Living mulches, such as turf, can protect the soil from absorbing too much of the sun's energy. Turf stands that are mowed at maximum height during the hottest part of the year can shade the soil and reduce its temperature. This practice is not always practical on a golf course. Allowing turf to stand a little taller, however, not only provides a natural shade for the soil, but also encourages a greater production of organic matter because more photosynthesized food is produced for root growth. Root systems can contribute from one-half to four tons of organic residues per acre per year. Rolling greens can help maintain ball speed during periods when the turf is mowed higher to moderate soil temperature. Syringing during the hottest part of the day can also help to lower soil temperatures and reduce plant stress.

Research has discovered that high soil temperatures are much more stressful to plants than high air temperatures. As the soil temperature rises, plants consume more of the energy that they photosynthesize. This phenomenon can reach the point where more energy is consumed than produced. At this stage, plants can deplete much of their energy reserves and become extremely stressed. Adequate amounts of soil organic matter can slow the progress of this syndrome. Soil organic matter holds more moisture and plants tend to grow finer root hairs in soil enriched with humus. Finer root hairs have a greater amount of surface area and can absorb more water than coarse root hairs in the same time period. The plant's ability to transpire is its mechanism to cool itself. Plants that are better able to regulate their temperature mitigate the effects of heat stress and consume less of their own energy. Net gains in energy result in greater production and contributions of organic residues for soil saprophytes, more carbon dioxide for plants, and more humus.

Water can be both friend and enemy to the existence of organic matter. Organisms that decompose organic residues are as dependent on moisture as are plants or any other living thing. Common sense dictates that moisture conditions that are ideal for plants are most often ideal for the accumulation of organic matter, simply because the production of organic residues is at an optimum level. If water is supplied through an irrigation system, it is important to monitor soil moisture levels carefully and to practice moderate watering techniques to encourage root growth. Research suggests that low volume, high frequency watering techniques can improve plant and soil health while using water more efficiently (Vargas, 1994). Other experts disagree and suggest that deep and infrequent watering is best for most plants. Water in excess of what is optimal for plants can often be counterproductive. Saturated soil has little room for oxygen, which is essential for healthy roots. Stress resulting from overwet soil can reduce plant vigor and the rate at which it produces organic residues. It can also contribute to severe compaction.

Over the years, different methods of determining soil moisture, such as moisture blocks, tensiometers, soil psychrometers, infrared thermometers, and time-domain reflectometry have been developed. Dissolved salts in the soil solution can affect some of these devices, resulting in false readings. Some are very expensive while others are simply impractical. The infrared thermometer, for example, doesn't measure soil moisture per se, but takes the temperature of plant leaves, which can be an indicator of moisture deficiency. A hotter than normal leaf means that not enough moisture is being transpired through the plant. The operator of this device must sample leaves in a number of different areas to determine average moisture needs.

A simpler method of calculating irrigation involves evapotranspiration (ET) data and it shows some promising results in terms of plant response and the reduction of stress-related problems. The application of water in amounts that replace approximately what is lost from both evaporation and transpiration seems to create optimal conditions for plant growth and hence, production of organic residues. Most modern irrigation equipment comes with computer software that calculates the amount of irrigation needed to replace ET loss. Factors that influence ET such as temperature, humidity, wind, and solar radiation are either manually entered into the computer or automatically recorded from weather monitoring equipment. The software can calculate water loss from ET and automatically govern the irrigation system to replace it. If a computer or ET software is unavailable, ET data can be obtained at local meteorological recording stations or your local turfgrass extension service (see also DTN in Sources and Resources). ET calculations can also be done manually. Meteorologists have a special pan they use to monitor evaporation. This equipment can be a little expensive but is available to anyone and can be set up in areas that mimic the average conditions on the course. A homemade system consisting of a shallow pan and a ruler is not as accurate or as easy to monitor but certainly more affordable. More pans can be used if ET data from specific areas on the course need to be monitored. The loss (or gain) of water from the pan is measured and multiplied by a crop coefficient to determine ET. Crop coefficients for turf usually range from 0.63 to 0.78 for warm-season grasses and from 0.79 to 0.82 for cool-season grasses (Christians, 1998). A more precise coefficient can usually be obtained from a local meteorologist or turfgrass extension agent. If, for example, water loss from the pan is one-fourth inch on a particular day and the crop coefficient for the species of grass on the course is 0.81, then:

$$0.25 \times 0.81 = 0.2025 \text{ inch of water}$$

In this particular case, two-tenths of an inch of irrigation will replace ET loss. The schedule at which ET loss is replaced should depend on soil texture, climate, environmental stress levels, and the superintendent's discretion. During certain periods of the season, it may be appropriated to replace ET loss on a daily basis or only once per week. ET calculations don't take into account surface or soil drainage, or other factors that would influence the decision of when, how often, and how much to irrigate.

The balance of air and water in the soil naturally regulates the accumulation of soil organic matter. Periods of excess water deplete the amount of air, inhibiting the activities of aerobic decay organisms. Saturated conditions for extended periods of time will eventually suffocate roots, which stresses the plant and reduces the production of organic residues. During dryer periods, air is abundant but moisture becomes the limiting factor. Maintaining adequate levels of soil organic matter can buffer the effects of too much or too little water.

This discussion of organic matter begs the question: What is an adequate level? The answer, unfortunately, is imprecise. As one begins to make important contributions of organic matter, such as applications of compost/sand topdress mixtures, it usually becomes apparent that levels prior to this action were inadequate. What is excessive is harder to define but chances are that level is economically impractical to attain. Each one-eighth inch of a topdress layer requires almost 17 yd^3 of material per acre. Burying turf with an excessively thick topdress layer is never sensible no matter what kind of material is being applied, but the expense of this action would suppress even the most overzealous tendencies. The USGA suggests that 0.5 to 2% organic matter in a green's soil is adequate. They also believe that levels in excess of that may lead to drainage problems. An analysis of soil extracted from the greens at coauthor R. Luff's course revealed organic matter levels at between 3 and 4% and drainage has never been a problem there. This is a surprisingly low figure after almost four decades of topdressing with compost but it represents an extremely functional ecosystem that consumes the energy in these organic residues to perform essential biological activities. Research from the University of Georgia showed that organic content in the surface layer can exceed 10% in less than a year if what they consider to be proper cultivation and topdressing are not carried out. We contend that any soil that accumulates that much dead plant matter in a single year must be nearly devoid of essential saprophytic soil organisms and that management should be focused more on cultivating a healthy and diverse biomass than on putting holes and sand in the turf. Excessive thatch and other undecomposed plant residues are not the kind of organic matter that is being advocated in this book. It is, however, more difficult for these materials to accumulate in a biologically active soil. Verticutting may still be necessary to remove some of this tough, woody material even in a soil with an active biomass but accumulation of these residues is much slower.

To prescribe a precise or ideal amount of soil organic matter for all courses everywhere does not take into consideration all the environmental factors that control its accumulation and its destruction (see Chapter 1, Soil Ecosystem). It is unlikely that many courses would have the resources to build organic matter levels far beyond optimum. The exception might be a course that was originally built on bog or muck soil and already has excessively high levels of soil organic matter. If the course has been in operation for some time without drainage problems, these high levels of organic matter may not be a problem.

Inadequate levels of soil organic matter and the beneficial soil organisms it supports can lead to a wide range of problems and an even wider range of symptoms. One of the most immediate problems is compaction, which inevitably leads to stress. Soil compaction causes plants to grow shorter, thicker roots that occupy a shallower portion of the soil and have less surface area. These stunted roots are unable to absorb adequate water and nutrients, resulting in lower shoot density, a reduction in carbohydrate reserves, and less drought resistance. There is more than a third less rhizome development in compacted soils, which limits the turf's ability to fill in thin areas. Water that is trapped in compacted soil cannot move up through soil horizons to the surface where it can evaporate and reduce the soil's temperature. Instead this water heats up from the increased thermal conductivity of a compacted soil and begins to poach the turf's roots. Where attention has been paid to the creation and preservation of organic matter, soils are less prone to compaction but it still occurs. Compaction is most often a result of compression from too much traffic in a given area. Traffic includes golfers on foot and in carts, maintenance vehicles, and mowers. Larger greens and tees can dilute the wear associated with a greater number of rounds played. Over time, greens and tees tend to shrink in size, usually from mowing practices. Most workers are reluctant to cut too close to the collar or the leading edge of the green for fear of scalping the perimeter and losing face or possibly losing their job. Consequently, and at an imperceptibly slow pace, fractions of inches accumulate into feet that eventually evolve into yards. Slowly but surely the shape and size of the greens and tees change. As they shrink, traffic is concentrated into a smaller area, which increases wear, tear, and compaction. As the number of rounds increase at a course, so should the size of the greens and tees. Scalping the collar may be unsightly for a few days but can help preserve the original size of the green. This procedure should only be done when turf is growing vigorously.

The weight of a golf cart with two or more riders and two or more golf bags can exert a fair amount of pressure on the soil. Multiply by the number of carts that travel the same route and the result is almost always compaction. Reducing compaction caused by cart traffic can be accomplished by designating certain paths that are paved, layered with gravel, or where turf quality is unimportant. This is often frustrating if golfers ignore the paths and repeatedly take other routes over more sensitive areas. Another strategy is to diffuse the traffic over a broader area and dilute the compaction. Scatter bars are 8 – 10 feet long rails mounted horizontally on the ground that reroute cart traffic to prevent a path of compaction from being created. These can be used where carts cross drainage ditches or streams, take shortcuts through perpetually wet areas, depart from tees, or just follow common but overused routes. Dispersing wear patterns should be done on a daily basis.

Walk-behind mowers can be used during wetter periods to reduce the compression

Scatter bars are used to diffuse traffic patterns. These bars are moved to different locations frequently to reduce compaction and wear. The greater the amount of play, the more often the bars are moved.

caused by heavier triplex mowers. These mowers aren't as fast or as popular with the help but the difference in weight is significant. The average walk-behind mower weighs less than 200 pounds whereas a triplex is somewhere near 1,200 pounds depending on the manufacturer and the choice of diesel versus gas engines. The triplex manufacturers try to minimize the impact caused by the weight of the machine by increasing the footprint and diluting the weight over a broader area. The Jacobsen people claim that their Greensking IV triplex greens mower has a footprint weight of 8.5 pounds per square inch (psi), which is impressive (if not incredible) for a machine that weighs 1,150 pounds. The walk-behind greens mower, however, riding on a 20-inch long by 8-inch diameter roller, is exerting less than half that weight per square inch. To put this in perspective, a 175 pound golfer with a size 10 shoe has a foot print weight of ~3.6 psi when walking (calculated with all his weight on one foot) and half that when standing.

Earthworms feed on clippings or compost topdress applications and their activities can alleviate compaction on a constant basis. They may or may not be able to keep up with the amount of traffic but it is reasonable to assume that the larger their population and the more resources available to them at or near the surface, the more tunneling and aeration they will perform. Their castings may be a nuisance but can easily be smoothed with a drag cable each morning. This process is called poling by some superintendents and not only does it dispatch earthworm castings but it removes dew and smoothes anomalies caused by other burrowing organisms.

Drag (or poling) cables vary in thickness but 150 feet seems to be about the right length for most greens. Coauthor R. Luff has two different cables. The first is a three-fourths inch thick, braided nylon cable with a lead-weighted core called a Fairway Snake (Engage Agro). It was specifically manufactured to remove the dew off fairways and reduce the turf's susceptibility to dollar spot. The other is a homemade rig consisting of a 300-foot long three-eighths inch steel rope, both ends of which are attached to an all-terrain-vehicle (ATV) via a widget that separates the two lengths by about two feet. At the other end, a 25-foot length of chain has been attached to keep the doubled cable

The Fairway Snake removes dew and smoothes surface anomalies such as earthworm casts or small ant mounds. This tool was designed for the fairway but co-author Luff uses it successfully on greens.

from twisting and to extend the sweeping area capacity. This doubled cable has been in use for over 30 years and seems to work as well if not better than the fancy new one. Poling begins by dragging the cable along the side of the green until the ATV is slightly beyond the farthest end. Then the ATV drives around two more sides of the green and straight away as the cable sweeps across the green. Trying to explain this on the page is a little like writing shoe-tying instructions. For the sake of clarity, let's say the driver is approaching the green from the south on an ATV dragging a 150-foot long poling cable. He drives along the eastern edge of the green until he is slightly beyond the northern edge. Then he turns, driving along the northern edge first and turns again to travel down the western edge. If the green is exceptionally large, he may have to follow the southern edge of the green for a short distance to sweep the entire area. It may be altogether im-

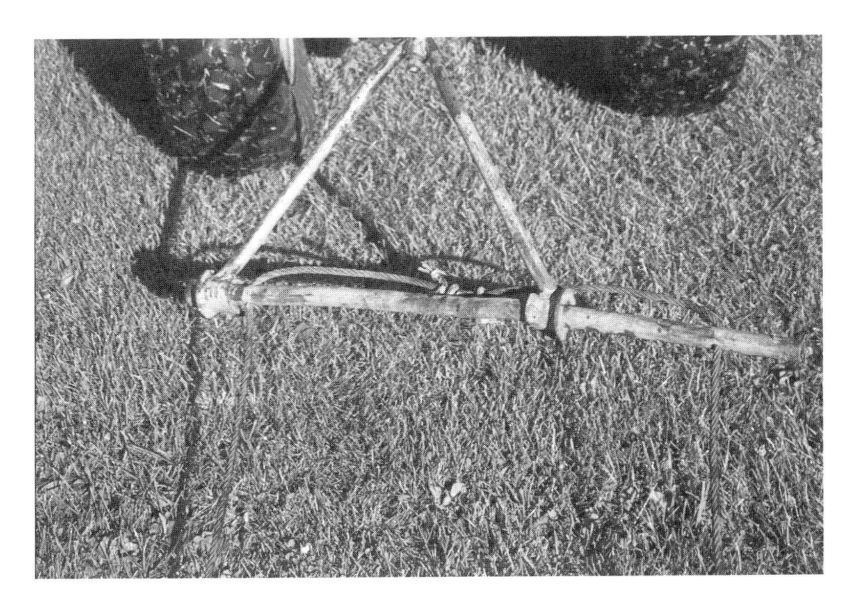

This widget serves as a connector for the original drag cable devised by R. DeWitt Luff in 1962.

The 150 foot long doubled cable sweeps the greens as well, if not better, than the Fairway Snake.

practical to drag some deeply bunkered greens. These greens may have to be whipped (or poled) by hand. Poling is a practice that takes an experienced operator very little time per green. It removes dew and mycelium, and smoothes earthworm castings, ant mounds, or debris left by green June beetles.

Earthworms can generate as much as 120 tons of castings per acre per year and this material contains valuable nutrients and other organisms that ultimately promote healthy, stress-resistant plants. Some ancient civilizations revered the earthworm to the extent that it became a protected species. Earthworms borrow down five feet or more—far deeper than any mechanical aeration tool—and return valuable minerals from soil depths to the surface. An acre of soil can be home to as many as 500,000 earthworms, building as much as 250 miles of tunnels in a week's time. Their activities not only im-

Earthworm casts are smoothed after a drag cable sweeps the green. The streaks disappear soon after the sun dries the surface. Earthworm casts are both fertile and biologically active.

prove the infiltration of oxygen and water to plant roots but also allow plant tops easier access to carbon dioxide the earthworms and other decay organisms generate. The resulting decrease in soil compaction at horizons beyond where aeration tools can reach allows roots easier access to deep soil moisture and nutrients. Earthworm tunnels are coated with a sticky mucous that preserves the passageway longer than those created by tines. Some scientists estimate that these tunnels can stay intact for as long as a year. This mucous is rich in nutrients and thought to contain hormones that stimulate plant growth. Turf roots often occupy these tunnels and can reach greater soil depths using less energy. When earthworms are active, thatch decomposition is relatively rapid. Earthworms reduce thatch layers more effectively than most other soil organisms combined. This is almost always a blessing but, on rare occasions, can be too much of a good thing. Thatch accumulation, especially on greens, can often lead to conditions that stress turf, cause turf decline, or provide opportunities for pest problems. Under these circumstances, thatch management by earthworms should be a welcome phenomenon. Thatch in small quantities, however, can be beneficial—mitigating compaction and other related problems—and in rare cases, excessive earthworm activity can reduce thatch below optimum levels. Earthworm activity normally subsides during hotter periods of the season, providing an opportunity for thatch to accumulate. Almost everyone has to deal with some thatch on greens even where earthworms are active. Regular verticutting during the heavy growth periods of the season can usually keep thatch under control without disrupting play but every so often a more aggressive approach is needed. Machines such as the Mataway or the Thatchaway reels cut deeply into the sod and remove a substantial amount of material including vital components of some weeds such as chickweed and clover. This practice is often more disruptive to the game than coring and is usually done at the end or the very beginning of the season. Some superintendents can control thatch by frequently topdressing. Strategic cup changing is another method of reducing thatch. Foot traffic has a significant impact on thatch and if it can be directed evenly throughout the green, thatch has a more difficult time accumulating. Sections of the green where cup placement is inappropriate usually have a much greater incidence of thatch. All of these thatch-reducing practices put together do not adequately replace the

Thatchaway reels cut deeply into the sod and remove a substantial amount of material including vital components of some weeds such as chickweed and clover

activities of earthworms. Their consumption of thatch is only one of the many benefits they can perform.

Earthworm activity can also reduce the likelihood of a condition know as localized dry spot (LDS) where the soil in small areas becomes extremely dry and resists rewetting. LDS is often associated with extremely sandy soil, excessive thatch, or compaction, all of which can be mitigated—directly or indirectly—by earthworms. The conventionally prescribed method of treating LDS is with wetting agents. Many wetting agents are natural and innocuous such as yucca extracts but, like pesticides, they often don't address the problem, just the symptom. Often, wetting agents will work well for awhile and then begin to fail because the severity of the problem has increased beyond the capability of the product. Another popular—but temporary—method of dealing with LDS is to perforate the soil's surface with a thin tine pitchfork or a water fork (which forces water through its tines) and then flood the area with water and wetting agents. Generally speaking, the cause of LDS is the deposition of organic hydrophobic substances on soil particles. Biological activity in a normal soil absorbs most of these substances, but in extremely sandy conditions bioremediation is often inadequate. Compost, clippings, and organic fertilizers stimulate the activities of earthworms as well as many other beneficial soil organisms that protect the soil from LDS. Sometimes LDS is a result of a topographical phenomenon such as high spots. The USGA recommends that high spots be hand-watered periodically to avoid LDS. Researchers at Rutgers suggest that it is possible to lower high spots by coring, then rolling without filling in the core holes. The space inside the holes often allows the ground to be compressed without actually causing compaction. These practices should not replace earthworm management but be used in conjunction with cultural methods that preserve this valuable resource.

Pesticides that kill earthworms inadvertently contribute to compaction and the loss of other benefits these organisms provide. Products that contain phorate, bendiocarb, carbaryl, chloropicrin, and benomyl are extremely toxic to earthworms. Other active ingredients that are less lethal but still very toxic to earthworms include ethoprophos, isazophos, methiocarb, propoxur, carbendazim, thiabendazole, and thiophanate-methyl (Ernst, 1995). The direct and indirect contributions earthworms offer to plant health are often sacrificed when any of these materials are applied.

Using mature compost in the topdress mix is not only important for earthworms but for most other beneficial soil organisms as well. The constant replenishment of composted residues feeds and reinoculates the valuable life forces in the soil, increases the water reserve, improves porosity (reducing compaction), promotes finer turf root hairs, reduces thermal conductivity, suppresses disease, regulates nutrient availability, and reduces stress. Topdressing with mature compost is one of the most important practices that an ecological superintendent can perform.

If topdressing is normally performed after core cultivation, the compost/sand mix can be applied whether cores are harvested or not. The amount of topdress material applied should be adjusted if cores are reincorporated. We recommend a compost/sand mix of 1:1 as a beginning ratio. The mix can be adjusted over time to better suit the uniqueness of the course, its regional characteristics, topography, and its soil. It is extremely important that well-made, well-aged compost is used (see Chapter 3, Compost). The compost will need to be screened and thoroughly mixed with sand before use. If screening and mixing equipment is beyond the club's resources to own and unavailable to rent, conventional sources of topdress mixes will likely have the necessary equipment. Once the mix is made it can be applied with a topdressing machine like any other topdress mix.

Care should be taken not to topdress during periods of severe stress. This precaution is advisable no matter what type of material is being used. The abrasiveness of sand being worked into the canopy of a stressed turf (not to mention the weight of the equipment) may create problems. Liquid applications of seaweed extract, humates, compost tea, or all three can be used to mitigate stressed turf. If the cores are to be reincorporated, less topdress material should be used to prevent smothering the canopy. Cores are usually broken apart with a drag mat, verticutter, or power brushes before the topdress material is applied.

Experiments at Rutgers suggest that the larger the diameter of the core extracted, the greater degree of compaction relief and the greater number of complaints from golfers. They arrived at five-eighths inch diameter core as a good compromise. Other researchers believe that the diameter of the core has less to do with adequate compaction relief than the percentage of surface area impacted. Statistically, greens where 15–20% of the surface is impacted, regardless of the tine size, fare the best. To calculate the percentage of surface impacted, multiply the tine diameter by one-half to determine the radius, then use the formula, $r^2 \times \pi$ (r = radius, π = 3.1416) to find the surface area impacted by a single core hole. Then multiply the surface area of one core hole by the number of cores per ft^2 and divide by 1.44. The answer equals the percentage of surface area impacted. If, for example, a superintendent uses a five-eighths (0.625) inch tine spaced two inches apart (36 holes per ft^2), then:

$$(0.625 \times 0.5)^2 \times 3.1416 \times 36 \div 1.44 \approx 7.67\%$$

To impact between 15 and 20% of the green's surface area, the aerator would have to pass over the green twice. This formula may be excessive for some greens and inadequate for others, depending on the amount of wear and the soil's resistance to compaction.

Experiments with solid tines or Hydrojecting® (Toro) showed more ephemeral relief but suggest that they may be a more prudent choice for midseason cultivation. Some managers avoid using solid tines because the tines glaze the walls of the hole, preventing penetration by roots, water, and air. This is less likely to occur in a soil that is almost all sand. The Hydroject tool causes less surface disruption than the solid tine, penetrates deeper, and does not glaze the walls of the hole it creates but does not replace the practice of core removal.

Like mowing, core cultivation is difficult to avoid on a golf course. Some superintendents topdress lightly and frequently as a substitute for coring but on high-traffic courses, this procedure may only be a postponement of the inevitable. Unlike mowing, coring disrupts the game. Not only is play often stopped for a period of time but golfers are usually unhappy about the operation for weeks to come. Unfortunately, coring is a necessary evil (to golfers) to alleviate compaction. Some believe that aeration can cause compaction beneath the depth of the tine's penetration and recommend that coring depth be altered from time to time. The tools that drill through the soil work slowly but do a good job of relieving compaction. They can penetrate more deeply than the typical tine.

Some superintendents like to measure compaction before treating it. Several different methods such as water infiltration, indicator weeds, knifing, or penetrometers can be employed. The water infiltration method can be as simple as observing how fast water soaks-in during irrigation or by using the tin can method. Both ends of the can are removed and one edge is sharpened with a grinder so that it can penetrate into the soil with minimal damage to the turf. A predetermined amount of water is poured into the can and soil absorption is timed. It is important to perform this test soon after aeration so that a comparison rate can be determined. Weeds that indicate compaction can be found in Table 5.2. This method is more appropriate for superintendents who are intimately familiar with the subtle changes that occur on their greens. The incidence of a single weed that tolerates compaction may be a false alarm. Knifing is the simple procedure of stabbing the ground with a pocketknife and observing the amount of force required to penetrate the soil. This method relies on a good memory as one must compare the required force soon after aeration to the resistance felt in subsequent tests. A penetrometer will give a numeric reading (usually in Newtons) that can be recorded and compared.

It is important to completely fill the core holes, especially in the spring to prevent insects such as sod webworm or cutworm adults from laying eggs in the partially filled holes. Once the topdress material has been worked into the canopy, a small amount of irrigation (~ one-tenth inch) can be used to wash the material through the canopy and reduce the stress of the operation. The use of liquid seaweed extracts or humates is often very helpful at this stage. The superintendent should notice that the turf makes a much quicker recovery with compost in the topdress material than it would with just sand or sand mixed with topsoil. He may also notice that it takes a longer period of time than usual for the green to become compacted again.

Zeolite (clinoptilolite) is another product that can be mixed with topdress materials to reduce compaction, improve water retention, and increase CEC. The stable structure of this natural material provides permanent pores that can retain as well as drain water. Research has shown significant benefits from using zeolite including greater porosity, increased resistance to compaction, better fertilizer efficiency, water conservation, reduced leaching, increased nitrogen efficiency, decreased localized dry spots, and improved nutrient utilization. Some researchers, however, have found that some of these products can hold water too tightly, making it less available to plants. Zeolite products are marketed under brand names such as Clino-lite,™ Eco-Sand,™ Agricolite,™ Zeoclere-30,™ Z-Plus,™ and ZeoTech,® to name a few, and all can be found on the World Wide Web.

Once the soil is compacted plants become stressed and weakened, giving greater opportunity to weeds, insects, and disease pathogens. Compaction, however, is not the only cause of stress. Plants experience stress from many different sources, some of which are very natural. Excessive heat, drought, cold, or precipitation can stress plants. Many different plant functions such as germination and flowering are also sources of stress. These natural events do not always overwhelm plants by themselves but coupled with the stress from golfers walking on, driving over (with carts), and scalping divots from turf makes stress management a bit more difficult for plants. Add to all of this the stress of maintenance. Almost everything we do to pamper turf causes stress. Turf stress is not difficult to mitigate but nearly impossible to avoid on a golf course.

Mowing, the not-so-surgical removal of plant tops is, without a doubt, a stressful practice and one that must occur daily. We are not suggesting that turf plants feel pain from their extremities being amputated but mowing is a type of injury. We all know how important it is to mow with sharp blades and bed knives because it minimizes the injury and reduces stress. Wear on bed knives is particularly severe after topdressing, especially if the topdress material is mostly sand. The more often topdressing is done, the more frequently back-lapping or grinding needs to be performed. Rotary mower blades should also be kept sharp even if they are only used in the rough. Damage to grass from dull rotary blades is unsightly and can lead to disease or other problems that may spread elsewhere on the course. There is talk of new mower technology that uses laser beams instead of blades to cut the grass. Not only is the cutting height extremely accurate but the wound is immediately cauterized by the beam and the clippings are dried and pulverized on the spot. A prototype of this machine, seen at a German equipment show (fall, 2000), also had leather seats, mobile internet access, and a CD player. (We're not kidding.) It hasn't been determined, as yet, if this new laser technology does, in fact, reduce turf stress but it certainly reduces fuel consumption, noise, and pollution. Unfortunately, at $30,000, many clubs may find it stressful to their budget (White, 2001a).

It has been said that we should not remove much more that one-third of the turf canopy when we mow. It is well documented that turf recovers just fine if more than a third is cut, but photosynthesis can be temporarily halted while reserve energy stored in the plant roots is used to grow new shoots. This causes stress too. Removing more than a third of the plants' tops can arrest root growth for as long as a month. If too much of the canopy is removed, photosynthesis slows to the point where plants must use nutrient reserves in their root systems to sustain themselves. When new top growth exposes itself to the sun, photosynthesis will again begin to produce nutrients for the roots but must first replenish the reserves that were depleted during the period when leaves were too short. The energy needed for normal functions plus the restocking of reserves can stop the growth of turf roots for almost a month. Loss of root growth during crucial times of the season can make the difference between turf with problems and turf that is relatively trouble-free, in both the short and long term. The degree of root growth is relative to how a plant fares through many different types of stresses such as heat, cold, and drought. Plus, the root's production of organic matter in the soil is a long-term asset that can benefit many future generations of turf plants.

Weather is often a main source of stress and a major contributor to turfgrass decline, and it is completely beyond anyone's control. An experienced superintendent, however, can often predict and prevent many weather-related problems before those conditions prevail. To do this, however, a certain degree of monitoring is necessary to compile and correlate necessary information. Some superintendents have their own weather monitoring equipment and keep track of degree-days, ET, humidity, etc., (Spectrum) while others get daily reports from DTN online (requires membership) or local meteorologists. All of this information is meaningless of course unless it correlates to impending problems. Researchers have found relationships between heating degree-days, for example, and certain insect or disease problems. Many of these maladies may not plague every superintendent but it's good to know the conditions under which the persistent problems exist. Books and articles written about turfgrass pests often contain information that matches the accumulation of certain weather conditions with the onset of certain pests. There are new high-tech computer systems that link to weather monitoring devices and contain software that automatically warns the user of possible impending problems based on weather data it has collected. Although these new systems have shown a great deal of promise, it is important that the superintendent keep a vigil. Commercial airlines have onboard computers that can literally fly the plane without assistance but they still hire and know the value of their pilots.

Sometimes the onset of a pest is ahead of the forecasted schedule because conditions on the course have accelerated susceptibility. Poor air movement, for example, can foster problems associated with leaf wetness well before overall moisture, humidity, or heat levels have reached a critical threshold. Increasing air movement, either by thinning obstacles or providing mechanical assistance with blowers, can decrease susceptibility associated with certain weather conditions.

One of the most difficult weather conditions for turf to deal with is heat and drought. Oppressively hot weather is a challenge for all living things but to face it without water is often a death-knell. The practice of syringing greens (i.e., a brief application of water) during periods of extreme heat is often all that is needed to prevent turfgrass decline. Many superintendents attempt to apply a specific amount of water (e.g., ~one-tenth inch) during the hottest part of the day, while others will go out with a hose equipped with a shower type nozzle and lightly water the green by hand. The effect of syringing is similar to what you might feel if, after sitting or working in the hot sun for a prolonged period of time, you were showered with cool water. This would not only provide instant relief because of the significantly lower water temperature, but also a sustained cooling from evaporation.

Extreme weather conditions cannot be controlled but can be tolerated by some varieties better than by others. Seed choices are important for many other reasons too. First and foremost is the suitability of the playing surface but a close second is the survivability of the selection. A tremendous amount of research has been done to determine the resilience of different varieties to disease, drought, traffic, low HOC, heat, cold, salt, insects, and excessive precipitation. This information is useful to a point but the savvy superintendent understands that just because a variety performs well in a university test plot doesn't always mean it will be the ideal choice for his course. Subtle differences in the climate, topography, and soil conditions may be just enough to negate most of what was learned by research. Additionally, irrigation, fertilization, and cultivation may also change conditions enough to render the test data relatively meaningless. The information gathered by scientists is important and a good starting point but the superintendent should probably conduct some trials of his own before committing to any specific varieties of seed.

Finding the right varieties is a huge step toward the reduction or elimination of pesticides. Species that thrive in a given area are often more competitive than most weeds and can be relatively resilient to damage from insects, pests, and disease pathogens. If poor soil conditions prevail, however, few varieties may be able to thrive. If a cultivar is found that performs relatively well in poor soil conditions and soil improvement becomes an objective down the road, that variety may no longer be the best choice. Changes in other conditions such as weather trends or exposure (e.g., shade reduction or expansion) may also make the original choice of seed inappropriate. Sometimes native, unnamed varieties may serve the superintendent best. Coauthor Richard Luff cultivates a nursery of native bentgrass that has greater resistance to dollar spot and other diseases than the conventional varieties he's tried. Seed varieties are always changing and new ones seem to appear every year. Seashore paspalum, for example, has become a popular warm-season alternative in the littoral regions of the country. Seed choices for different latitudes or altitudes; proximity to fresh or salt water; arid, semiarid, temperate, tropical, or subtropical zones; and different soil conditions would be inappropriate to discuss here. Not only would it take hundreds of pages but the information is quite ephemeral. What is new as of this writing may be old news by the time this book is printed.

The turfgrass extension system of the local land-grant university is usually a good source of information on new seed varieties and the conditions under which they thrive. Many large universities conduct annual trials and publish their results. These data usu-

Natural areas as described in Chapters five and six. These areas are not seeded or planted. The plants are natural indigenous species that thrive on neglect.

ally describe performance based on many different criteria, including disease resistance, drought, heat, shade and wear tolerance, thatch, sod strength, performance at different cutting heights, ball speed, response to environmental stresses, and appearance. Another source of information is the National Turfgrass Evaluation Program (NTEP). NTEP collects data from local, regional, and national sources and stores them on their own database. NTEP can be accessed on the web at www.ntep.org.

Some areas on the course are often more trouble than they are worth. They seem to require an inordinate amount of maintenance, are susceptible to every kind of problem, and are not strategic areas in terms of the game. These spots are perfect candidates for intentional neglect. Let Mother Nature grow what she wants.

The addition of natural areas in certain portions of the rough or out-of-bound areas contributes not only to the beauty of the course and the reduction of maintenance but also provides habitat for many different beneficial insects, arthropods, birds, mammals, and soil organisms. Natural areas can contain wildflowers, ornamental grasses, perennials, or simply the indigenous species of plants that thrive when the area is undisturbed. Natural areas are ideal in perpetually wet spots where irises, cattails, and other wetland plants can thrive. Turf growing in wet areas is usually prone to problems and requires more intensive management. Some managers believe that natural areas can contribute to the proliferation and spread of noxious weed seeds but those plants that thrive in natural areas don't usually compete well at normal cutting heights in the fairway or even the rough. Some managers like to burn natural areas in the late fall or early spring to refresh

Natural areas as described in Chapters five and six. These areas are not seeded or planted. The plants are natural indigenous species that thrive on neglect.

the area and kill weed seed, pathogens, and dormant insects. Unfortunately, there's no guarantee that what is being killed is in any way harmful or troublesome. More often than not, securing the necessary permits from the fire department is more trouble than it is worth. If tall plant residues pose a visual problem, they can easily be dispatched with a string trimmer and the detritus can be added to the compost heap. Natural areas are generally mowed down at least once per year to prevent establishment of unwanted trees or shrubs.

Some managers like to remove the leftover debris after mowing down natural areas but that removal is like harvesting a crop. The nutrients extracted from the soil by those plants are not being returned and ultimately need to be replaced. The same is true for leaves that fall from the trees scattered around the course. To avoid obstruction of the game during the fall, leaves need to be blown off the greens but, like clippings, they can be shredded by a rotary-type rough mower and left to replenish the soil.

Albeit blasphemous, if the superintendent's objective is the reduction or elimination of pesticides, a discussion of artificial turf is appropriate. Advancements in synthetic turf technology have reached a point where playing surfaces are comparable to natural turf that has been managed to the nth degree (White, 2001b). In fact, some manufacturers (Player's Turf International LLC, Field Turf, Inc.) boast balls stimping up to 15 feet. Their technology combines a thick mat of green, synthetic shag and special topdressing materials that, in varying amounts, create different surfaces. The conditions of these simulated surfaces range from roughs to greens and also include athletic fields, tennis

Out of bounds markers protect a killdeer (American plover) nest on the fairway. Eco-conscious maintenance involves protecting wildlife even if they don't directly contribute to the welfare of the course.

courts, or just maintenance-free lawns. These playing surfaces never need mowing, aeration, or pest control and can last 20 years or longer. If pesticide elimination becomes mandatory and few if any of the alternatives or practices mentioned in this book are practical, it might be worth considering artificial turf. It isn't a lot less natural than some of the management techniques we currently use.

Chapter 7

Taking Responsibility

Coauthor Richard Luff's father, Peter, built the Sagamore Hampton Golf Club in 1962, and without any pressure from environmental groups, decided that construction and maintenance of this course could be and would be implemented without resorting to chemical management. The only reason it worked was because he wanted it to.

Organic or Natural Method Maintenance, as coined by Peter Luff, was as foreign a turf maintenance term in 1962 as Hydrojecting. Clearly 30 years ahead of his time, the construction and maintenance methods Peter practiced on his 18-hole golf course influenced hundreds in the turf industry. From well-known superintendents to homeowners interested in maintaining their lawn naturally, all have learned from the Natural Method Maintenance program. Modest to a fault, when told he was on the cutting edge, Peter would chuckle and say that what he practiced was 40 to 50 years behind the times.

Peter's techniques originated from the forgotten basics of turf maintenance practiced in the 1920s, 1930s, and 1940s. The post-World-War-II chemical age dramatically changed many aspects of golf course maintenance. Peter's philosophy was a reaction to his early years on a grounds crew. During the mid to late 1940s, while working at the family-operated Sagamore Spring Golf Club in Lynnfield, Massachusetts, Peter experienced the good and bad aspects of turf maintenance. Years later, Peter would often relate stories of innocence and ignorance surrounding early chemical use on the course. "While hand-spraying fungicide on hot days," he remembered, "members of the crew would spray each other to 'cool down,' never considering the potential for harm. Other times the crew would mix raw mercury with sand in a closed building without the benefit of protective clothing." These experiences raised questions in Peter's mind about what they were doing and why they were doing it.

Peter's father and grandfather explored methods of organic gardening and composting. Their experiences inspired him to explore the possibility of applying these same principles to turf maintenance. Peter would often use J.R. Rodale's *Encyclopedia of Organic Gardening* as a reference for Natural Method Maintenance. He sensed that the basic organic tenets espoused in this tome could, with a bit of modification, be applicable to the golf course. Peter quickly realized that working with Mother Nature, instead of against her, proved fruitful and rewarding. Peter reasoned that healthy turf would follow healthy soil, which could only be created by promoting an abundance of soil organisms as an integral part of the maintenance program. He understood that soil life was the key to soil improvement.

In 1960 opportunity knocked, and he had a chance to practice what he preached. The family business expanded with the purchase of a 380-acre parcel of land in North

Hampton, New Hampshire, upon which an 18-hole golf course was to be built. Peter said he would take on the monumental task of building and managing the course on the condition that he could maintain the course by Natural Method Maintenance. The skeptics were plentiful but Peter demonstrated the strong relationship between resolve and outcome. Capable of working 16-hour days for months on end and being extremely resourceful, an 18-hole golf course was built on time and under budget. When reflecting on building the course, Peter reworded the phrase, "Necessity is the mother of invention" to a more germane "Lack of funds is the mother of invention."

Not only did fiscal restraints dictate creative construction methods, but the land had kame and kettle features that favored a unique course design. Leaving these features intact despite their valuable gravel content also went hand in hand with the Natural Method Maintenance philosophy. Peter developed what he liked to call *flowerpot greens*, which was a construction method inspired by severe fiscal restraints. The greens were located and contoured to take advantage of the existing drainage patterns. Once the contours were established the area was covered with 4 inches of sharp sand and topped with eight inches of a 30% loam, 70% sand mixture. Approximately 300 yards of sand and 60 yards of loam were used per green. Only 40 acres of forestland were cleared, as most of the fairways were formed from old pastureland. Only the tees and greens were seeded. Peter was able to transform the rough pasture into acceptable turf simply by mowing and minimal fertilization. In the areas that were seeded, Peter believed that the broadest possible spectrum of seed varieties should be applied to facilitate the process of natural selection.

Fertilization is common to all golf courses and necessary to maintain competitive conditions. While Peter limited fertilization to greens, tees, and spot applications on fairways, the products he used were anything but common. In the 1960s and 1970s, *natural organic* fertilizers amounted to Milorganite and a few other hard to find and harder to spread products. Peter didn't believe that a balanced fertilizer program could be achieved using contemporary mainstream products so he was continually searching for products that would feed soil organisms first and foremost. As a result, the local feed stores received most of Peter's business. Fertilizers used in the Natural Method Maintenance program included soybean meal, castor pomace, dried blood, granite dust, cottonseed meal, linseed meal, cotton hull ash, greensand, sunflower hull ash, leather dust, crushed basalt rock, corn gluten meal, langbeinite dust, kelp meal, hardwood ash, and seaweed extract. While some of these products were readily available, many were not. Peter would inquire throughout the United States and Canada during the off-season for different natural products. One particular order of dried blood, bone meal, and hog's hair, delivered via the local post office, raised some suspicions among the locals who imagined that Peter was up to something other than just building a golf course.

The fertility program often involved the whole family. The course's proximity to the shore made it possible to harvest seaweed and kelp for the compost piles. The Luff family's day at the beach not only involved sun and surf but mandatory seaweed collection as well.

Peter had a unique situation. He was able to call the shots and answered only to Mother Nature—and the customers of course. Maintaining competitive conditions was paramount to the survival of the business but he understood that not being held accountable to a greens committee or a vocal membership was a luxury that was crucial to what he was trying to accomplish. "Untie the hands of the superintendent" he would say when discussing the pressures faced by many superintendents. He encouraged every superintendent who contacted him to take responsibility for their maintenance program and not be so easily influenced by salesmen, golfers, industry experts, trade journals, or the golf course down the street. In a field where independent thought was so uncommon, Peter led the way and tried to impress all those who were interested to "follow nature's lead and listen to her every day."

Peter described himself as a conservationist. He had a love affair with nature and was never as interested in golf. He looked to Mother Nature for answers to all of his questions about maintenance, pestilence, and life in general. He took pleasure in every natural aspect of the golf course from the feeding activity of starlings and flickers to the tiniest snow fleas. "The game of golf should be a healthy outdoor experience; it should expose one to the elements and to nature. That's what it should be, not a source of pollution." Peter would say, "We are used to thinking of nature as the enemy but *in order to command nature, you must learn to obey her.*" *(Sir Francis Bacon).*

Peter passed away in 1998. He left behind an ecological legacy that lives on through those who practice what he preached. His pioneering spirit inspired many, not just in the golf course industry, to take responsibility for their environment even if others won't, and to do what they can to preserve and promote the living component of the ecosystem.

Glossary

This glossary is intended to provide definitions for terms used within the text of this book and some of the terms used in many of the sources listed in the Sources and Resources section.

Abiotic – Not living.

Acid – Any substance that can release hydrogen ions in a solution.

Actinomycetes – Decay microorganisms that have a fungus-like appearance but, like bacteria, do not contain a well-defined nucleus.

Adsorb – See Adsorption.

Adsorption – The adherence of one material to the surface of another via electromagnetic forces, e.g., dust to a television screen.

Adventitious rooting – Roots emanating from aboveground plant parts.

Aerobic – Needing oxygen to live.

Alkaline – Refers to substances with a pH greater than 7.

Allelochemicals (or **Allelopathic chemicals**) – Refers to substances produced by one plant that have a negative effect on another.

Allelopathic – Usually refers to the negative influence a plant has on other plants or microorganisms.

Amorphous – Without consistency in its structure or form.

Anaerobic – Refers to an environment with little or no oxygen or organisms that require little or no oxygen to live.

Anaerobiosis – Life in the absence of oxygen.

Antagonist – Any organism that works against the action of another.

Anthropogenic – Caused by human action.

Apatite – A natural phosphate mineral.

Aragonite – Calcium carbonate (lime) formed by shellfish.

Arthropod – Refers to a group of animals with segmented bodies and exoskeletal structure such as insects, spiders, and crustaceans.

Assimilation – Digestion and diffusion of nutrients by an organism for growth and/or sustenance.

ATM – Atmosphere, a measurement of pressure of a gas.

Atmosphere – Refers to the naturally existing gases of any given environment. Also a measurement of pressure (see ATM).

Autoclave – A machine that, with heat and pressure, can react gases with other materials. It is also used to sterilize tools and equipment.

Autotrophs – Organisms that can synthesize organic carbon compounds from atmospheric carbon dioxide, using energy from light or chemical reactions.

Base cation – A positively charged ion, historically belonging to the earth metal family (potassium, magnesium, calcium, etc.).

Bases – See Base cation.

Biomass – The cumulative mass of all living things in a given environment.

Biota – Biomass.

Biotic – Pertaining to life or living organisms.

Botany – The study of plants.

Calcite – Calcium carbonate (lime).

Carbohydrates – A group of organic compounds that include sugars, starch, and cellulose.

Carbon : Nitrogen ratio – A ratio measured by weight of the number of parts carbon to each part nitrogen, e.g., 10:1, 50:1, etc.

Carnivores – Organisms that consume animals or insects for sustenance.

Cation – An ion of an element or compound with a positive electromagnetic charge.

Cation exchange capacity – The total amount of exchangeable cations that a given soil can adsorb.

Cellulose – The most abundant organic compound on earth, found mostly in the cell walls of plants.

Chelation – The combination of metal (inorganic) and organic ions into a stable compound sometimes referred to as a chelate.

Chlorosis – Loss of normal green color in plants.

Colloids – Very small soil particles with a negative electromagnetic charge capable of attracting, holding, and exchanging cations.

Consumer – See Heterotrophs.

Cultivar – Refers to a specific plant variety.

Detritus – The detached fragments of any structure, whether biotic or abiotic, that are decomposing or weathering.

Dolomite – Calcium, magnesium carbonate (magnesium lime).

Edaphic – Refers to factors, such as soil structure, atmosphere, fertility, and biological diversity, that influence the growth of plants.

Edaphology – The study of edaphic factors.

Edaphos – Greek word meaning soil.

Entomo – Prefix pertaining to insects.

Enzymes – A group of proteins that hasten biochemical reactions in both living and dead organisms.

Eutrophy – Refers to the excessive nutrient enrichment of ponds or lakes, causing the accelerated growth of plants and microorganism and depletion of oxygen.

Evapotranspiration – Refers to water loss from the soil from both evaporation and transpiration through plants.

Exude – The release of substances from cells or organs of an organism.

Faunal – Pertaining to microscopic or visible animals.

Fecundity – Refers to the reproductive capabilities of an organism.

Floral – Pertaining to plants or bacteria, fungi, actinomycetes, etc.

Free oxygen – Gaseous oxygen not bound to other elements.

Furrow slice – Plow depth of approximately six to seven inches.

Geoponic – Pertaining to agriculture or the growing of plants on land.

Gustation – Or gustatory refers to an organism's sense of taste.

Hemicellulose – A carbohydrate resembling cellulose but more soluble; found in the cell walls of plants.

Herbivores – Organisms that consume plants for sustenance.

Heterotrophs – Organisms that derive nutrients for growth and sustenance from organic carbon compounds but are incapable of synthesizing carbon compounds from atmospheric carbon dioxide.

Humification – The biological process of converting organic matter into humic substances.

Humology – The study of humus.

Hydrolysis – The reaction of hydrogen (H) or hydroxyl (OH) ions from water with other molecules, usually resulting in simpler molecules that are more easily assimilated by organisms.

Hyphae – A microscopic tube that is a basic component of most fungi in their growth phase.

In situ – Refers to natural or original position. Example: organisms in situ may respond differently to a stimulus than they would in a laboratory.

Ion – Any atom or molecule with either a positive or negative electromagnetic charge.

Kairomone – A chemical substance produced by organisms (e.g., plants or insects) that attracts another species or the opposite gender to it. Example: 1) Fruit produces kairomones that attract certain insects. 2) Many insects produce kairomones called pheromones that attract the opposite sex.

Kame – A ridge, hill, or mound deposited by glacial melt water.

Kettle – A hollow formed by glacial drift.

Ligand – A compound, molecule, or atom with the capability of bonding with another compound, molecule, or atom.

Lignin – A biologically resistant fibrous organic compound deposited in the cell walls of cellulose whose purpose is strength and support of stems, branches, roots, etc.

Lodging – When plants become too top-heavy to stand upright and instead lie over onto the ground.

Macro – A prefix meaning large.

Meso – A prefix meaning middle.

Metabolism – The biological and chemical changes that occur in living organisms or the changes that occur to organic compounds during assimilation by another organism.

Metabolite – A product of metabolism or a substance involved in metabolism.

Meteorology – The study or science of the earth's atmosphere.

Methodology – A system, or the study of methods.

Micelle – (Micro-cell) A negatively charged (colloidal) soil particle most commonly found in either a mineral form (i.e., clay) or organic form (i.e., humus).

Micro – A prefix meaning small, usually microscopic.

Mineralized – The biological process of transforming organic compounds into nonorganic compounds (minerals), e.g., mineralization of protein into ammonium.

Mineralogy – The study or science of minerals.

Mitigate – To lessen.

Mmho or Millimho – A thousandth of a mho, which is a measure of a material's ability to conduct electricity. Usually used in soil tests to determine salt levels.

Molecule – The smallest particle of a compound that can exist independently without changing its original chemical properties.

Monoculture – The cultural practice of growing only one variety of crop in a specific area every season without variance.

Morphology – The study or science of the form or structure of living organisms.

Mucilage – Compounds synthesized by plants and microbes that swell in water, taking on a gelatinous consistency, that function to maintain a moist environment and bind soil particles together to form an aggregation.

Myco – A prefix that refers to fungi.

Nitrification – A process performed by soil bacteria that transforms ammonium nitrogen into nitrite and, finally, nitrate nitrogen. Nitrate is the form of nitrogen most often used by plants.

OM – Abbreviation for organic matter.

Oxidation – Usually refers to the addition or combination of oxygen to other elements or compounds.

Oxidize – To add oxygen. See Oxidation.

Parent Material – The original rock from which a soil is derived.

Pedology – The study or science of soils.

Pedosphere – The top layer of the earth's crust, where soils exist.

Phenology – The study or science of biological phenomena and their relationship to environmental factors.

Pheromone – A chemical produced by an insect or other animal that attracts another member of the same species, usually of the opposite sex.

Phyllosphere – Leaf surfaces.

Physiology – The study or science of the biological functions and/or activities of living organisms.

Phyto – A prefix referring to plants.

Phytopathogen – A plant parasite.

Phytotoxic – A substance that is toxic to plants.

Porosity – Refers to the spaces between soil particles.

Producer – See Autotrophs.

Rhizosphere – The area of soil in immediate proximity to roots or root hairs of plants.

Saprophyte – An organism that can absorb nutrient from dead organic matter.

Senescence – The aging process.

SOM – Abbreviation for soil organic matter.

Steward – A person who manages or cares for property of another. In agriculture, the term can refer to someone who cares for his own land but believes that ownership does not entitle one to dispose of the soil's resources for his/her own personal gain.

Substrate – Material used by microorganisms for food.

Superfluity – Oversupply.

Symbiotic – A relationship between two organisms, usually obligatory and often of mutual benefit.

Synergy – Where the activities or reactions of two or more organisms or substances are greater than the sum of the agents acting separately.

Taxonomy – The science of classification.

Tectonic – Pertaining to the structure and form of the earth's crust.

Texture analysis – An analysis of soil particles determining the percentages of sand, silt, and clay.

Throughfall – Moisture or precipitation that drips from aboveground plants, such as trees, to the ground. Throughfall is thought to contain some substances leached from leaf surfaces.

Topography – Pertaining to the specific surface characteristics of a given landscape.

Trophic levels – Levels of consumers within a food chain in relation to producers of organic nutrients, such as plants. For example, producers—primary consumers—secondary consumers—tertiary consumers—decay organisms.

Valence – A measurement of how many electrons an atom or molecule can share in a chemical combination. A positive valence indicates electrons offered in a chemical bond, whereas a negative valence is the number of electrons that can be accepted.

Volatile – Refers to substances that can easily change, often into a gas.

Sources and Resources

Abernathy, S.D., R.H. White, P.F. Colbaugh, M.C. Engelke, G.R. Taylor, and T.C. Hale. 2001. Dollar Spot Resistance Among Blends of Creeping Bentgrass Cultivars, *Crop Science* 41:806–809, Crop Science Society of America, Madison, WI.

Abu-Hamdeh, N.H. and R.C. Reeder. 2000. Soil Thermal Conductivity: Effects of Density, Moisture, Salt Concentration, and Organic Matter, *Soil Science Society of America Journal*, 64:1285–1290, Soil Science Society of America, Madison, WI.

Acadian Seaplants Limited, 30 Brown Ave., Dartmouth, Nova Scotia, Canada, B3B 1X8, phone: 902-468-2840, fax: 902-468-3474, www.acadianseaplants.com, customerservice@acadian.ca.

Adams, J., Ed. 1992. *Insect Potpourri: Adventures in Entomology*, Sandhill Crane Press, Inc., Gainesville, FL.

Adams, W.A. and R.J. Gibbs. 1994. *Natural Turf for Sport and Amenity: Science and Practice*, CAB International, Wallingford, United Kingdom.

Albrecht, W.A. 1938. Loss of Organic Matter and Its Restoration, U.S. Dept. of Agriculture Yearbook 1938, pp. 347–376.

Alm, S.R. 1994. Oriental Beetles in New England, *Turf Notes*, Volume 4, No.5, UMass Extension, Worcester, MA.

Alm, S.R. 1995. Annual Bluegrass Weevil, Hyperodes (Listronotus), *Turf Notes*, Volume 5, No. 2, UMass Extension, Worcester, MA.

Alm, S.R. 1999. Turfgrass Insect IPM—Present Status and Future Opportunities, *Turf Notes*, Volume 8, No. 2, UMass Extension, Worcester, MA.

A2LA, American Association for Laboratory Accreditation, 5301 Buckeystown Pike, Ste. 350, Frederick, MD 21704, phone: 301-644-3248, fax: 301-662-2974.

Anderson, A.B. 2000. Science in Agriculture: Advanced Methods for Sustainable Farming, *Acres USA,* Austin, TX.

Arshad, M.A. and G.M. Coen. 1992. Characterization of Soil Quality: Physical and Chemical Criteria, *American Journal of Alternative Agriculture*, Volume 7, Nos. 1 and 2, pp. 25–31. Institute for Alternative Agriculture, Greenbelt, MD.

ASA# 47. 1979. *Microbial—Plant Interactions,* American Society of Agronomy, Madison, WI.

Ash, S. A. 1998. Golf Course IPM for English Daisy, *IPM Practitioner*, Volume 20, No. 9, Bio-Integral Resource Center, Berkeley, CA.

ATTRA (Appropriate Technology Transfer for Rural Areas), P.O. Box 3657, Fayetteville, AR 72702, phone: 800-346-9140, web address: www.attra.org.

Autrusa Compost Consulting (George Leidig), 941 Perkiomonville Rd., Perkiomonville, PA 18074-9607, phone: 610-754-1110, e-mail: autrusa@aol.com, website: www.autrusa.com

Ayad, J.Y., J.E. Maham, V.G. Allen, and C.P. Brown. Effect of Seaweed Extract and the Endophyte in Tall Fescue on Superoxide Dismutase, Glutathione Reductase, and Ascorbate Peroxidase Under Varying Levels of Moisture Stress, T4.11.1 & EF4.5.1, Dept. of Plant and Soil Science, Texas Tech. Univ., Lubbock, TX.

Baker, R.R. and P.E. Dunn, Eds. 1990. *New Directions in Biological Control,* Alan R. Liss, Inc., New York, NY.

Balogh, J.C. and W. J. Walker. 1992. *Golf Course Management & Construction*, Lewis Publishers, Boca Raton, FL.

Barbosa, P. and D.K. Letourneau, Eds. 1988. *Novel Aspects of Insect—Plant Interactions,* John Wiley & Sons, New York, NY.

Barbosa, P., V.A. Krischik, and C.G. Jones, Eds. 1991. *Microbial Mediation of Plant—Herbivore Interactions.* John Wiley & Sons, Inc., New York, NY.

BBC Laboratories, Inc., 1217 North Stadem Drive, Tempe, AZ 85281, phone: 480-967-5931, fax: 480-967-5036, website: www.bbc-labs.com

Beard, J.B. 1995. Mowing Practices for Conserving Water, *Grounds Maintenance*, January 1995. Intertec Publishing, Overland Park, KS.

Beck, M. 1997. The Secret Life of Compost, *Acres USA*, Austin, TX.

Beck, M. 2000. More Secrets of Compost, *Acres USA*, February, 2000, Austin, TX.

Bhowmik, P.C., R.J. Cooper, M.C. Owen, G. Schumann, P. Vittum, and R. Wick. 1994. *Professional Turfgrass Management Guide.* University of Massachusetts Cooperative Extension System. Amherst, MA.

Bigelow, C.A., D.C. Bowman, D.K. Cassel, and T.W. Rufty Jr. 2001. Creeping Bentgrass Response to Inorganic Soil Amendments and Mechanically Induced Subsurface Drainage and Aeration, *Crop Science* 41:797–805, Crop Science Society of America, Madison, WI.

BioSafe Systems, P.O. Box 936, Glastonbury, CT 06033, phone: 888-273-3088, www.BioSafeSystems.com.

Bormann, F.H., D. Balmori, and G.T. Geballe. 1993. *Redesigning the American Lawn: A Search for Environmental Harmony*, Yale University Press, New Haven, CT.

Bosworth, S. Using Plants as Indicators for Diagnosing Soil and Turf Problems, unpublished, available from Department of Plant and Soil Science, University of Vermont, Burlington, VT.

Briggs, S.A. and N. Erwin, 1991. *Pesticides and Lawns,* Rachel Carson Council, Inc., Chevy Chase, MD.

Brady, N. C. 1974. *The Nature and Properties of Soils*, Macmillan Publishing Co. Inc., New York, NY.

Brame, B. 1996. Ergonomic Tee Divot Filling, *USGA Green Section Record*, Volume 34, No. 3, USGA, Far Hills, NJ.

Brede, D. 2000. *Turfgrass Maintenance Reduction Handbook: Sports, Lawns, and Golf*, Ann Arbor Press, Chelsea, MI.

Brookside Laboratories, Inc., 308 Main Street, New Knoxville, OH 45871, phone: 419-753-2448, fax: 419-753-2949, e-mail: SHANER@BLINC.COM (lab manager).

Buhler, D.D. Ed. 1999. *Expanding the Context or Weed Management,* The Haworth Press, Inc., New York, NY.

Bunce, R.G.H., L. Ryszkowski, and M.G. Paoletti. 1993. *Landscape Ecology and Agroecosystems,* Lewis Publishers, Boca Raton, FL.

Callahan, P.S. 1975. *Tuning in to Nature: Solar Energy, Infrared Radiation, and the Insect Communication System,* The Devin-Adair Co., Old Greenwich, CT.

Caron, C. 2000. Regenerating Soils with Ramial Chipped Wood, *The Ecological Landscaper,* Volume 7, No.1, Ecological Landscaping Association, Framingham, MA.

Carpenter-Boggs, L., A.C. Kennedy, and J.P. Reganold. 2000. Organic and Biodynamic Management: Effects on Soil Biology, *Soil Science Society of America Journal*, 64:1651–1659, Soil Science Society of America, Madison, WI.

Carrow, R.N. 1994. Understanding and Using Canopy Temperatures, *Grounds Maintenance*, May 1994. Intertec Publishing, Overland Park, KS.

Casagrande, R.A. 1993. *Sustainable Sod Production for the Northeast*, Department of Plant Sciences, University of Rhode Island, Kingston, RI.

Chenu, C., Y. Le Bissonnais, and D. Arrouays. 2000. Organic Matter Influence on Clay Wettability and Soil Aggregate Stability, *Soil Science Society of America Journal*, 64:1479–1486, Soil Science Society of America, Madison, WI.
Cherim, M.S. 1994. The General Principals of Biological Pest Control, *The Plantsman*, June/July 1994. New Hampshire Plant Growers Association, c/o UNH, Durham, NH.
Cherim, M.S. 1998. *The Green Methods Manual: The Original Bio-control Primer*, 4th ed., The Green Spot, Ltd., Publishing Division, Nottingham, NH.
Chet, I., Ed. 1987. *Innovative Approaches to Plant Disease Control*, John Wiley & Sons, Inc., New York, NY.
Christians, N. 1998. *Fundamentals of Turfgrass Management*, Ann Arbor Press, Chelsea, MI.
Cockfield, S.D. and D.A. Potter. 1984. Predation of Sod Webworm (Lepidoptera: Pyralidae) Eggs as Affected by Chlorpyrifos Application to Kentucky Bluegrass Turf, *Journal of Economic Entomology*, Volume 77, No. 6.
Coelho, R.W., J.H. Fike, R.E. Schmidt, X. Zhang, V.G Allen, and J.P. Fontenot. 1996. Influence of Seaweed Extract on Growth, Chemical Composition, and Superoxide Dismutase Activity in Tall Fescue, Dept. of Crop and Soil Environmental Sciences, Virginia Polytechnic Institute and State University, Blacksburg, VA.
Colbaugh, P.F. 1994. Dealing with Algae, *Grounds Maintenance*, May 1994. Intertec Publishing, Overland Park, KS.
Connellan, W. 1921. Compost and the Construction of Compost Heaps, *USGA Green Section Record,* USGA, Volume 1, No. 4, Far Hills, NJ.
Couch, G. 2000. Black Cutworm IPM, *Turf Magazine*, March 2000, Moose River Publishing, LLC, St. Johnsbury, VT.
Couch, G. 2000. Does Milky Spore Disease Work?, *Cornell University Turfgrass Times*, Volume 11, No. 3, Cornell University, Ithaca, NY.
Couillard, A., A.J. Turgreon, and P.E. Rieke. 1997. New Insights into Thatch Biodegradation, *ITRJ,* 8:427–435.
Craul, P.J. 1992. *Urban Soil in Landscape Design*, John Wiley and Sons, Inc., New York, NY.
Cummings, J.L., J.R. Mason, D.L. Otis, and J.F. Heisterberg. 1991. Evaluation of Dimethyl and Methyl Anthranilate as a Canada Goose Repellent on Grass, *Wildlife Society Bulletin* 19:184–190.
Dalthorp, D., J. Nyrop, and M.G. Villani. 2000. Foundations of Spatial Ecology: The Reification of Patches Through Quantitative Description of Patterns and Pattern Repetition, *Entomologia Experimentalis et Applicata*, 96:119–127, Kluwer Academic Publishers, Netherlands.
Dalthorp, D., J. Nyrop, and M.G. Villani. 2000. Spatial Ecology of the Japanese Beetle, *Popillia japoinca, Entomologia Experimentalis et Applicata*, 96:129–139, Kluwer Academic Publishers, Netherlands.
Dannegerger, K. 2000. Thermal Tolerance: The Role of Heat Shock Proteins, *Turfgrass Trends,* Volume 8, No. 7, Washington, DC.
Davidson, R.H. and W.F. Lyon. 1987. *Insect Pests of Farm, Garden and Orchard*: 8th ed., John Wiley and Sons, Inc., New York, NY.
Deal, E.E. 1967. Mowing Heights for Kentucky Bluegrass Turf, *Agronomy Abstracts,* Volume 59, No. 150, Nov. 1967, American Society of Agronomy, Madison, WI.
De Ceuster, T. J.J. and H.A.J. Hoitink. 1999. Prospects for Composts and Biocontrol Agents as Substitutes for Methyl Bromide in Biological Control of Plant Diseases, *Compost Science and Utilization,* Volume 7, No. 3.
De Ceuster, T. J.J. and H. A.J. Hoitink. 1999. Using Compost to Control Plant Diseases, *Biocycle,* The Journal of Composting and Recycling, Volume 40, No. 6, Emmaus, PA.
Dernoeden, P.H., M.J. Carroll, and J.M. Krouse. 1993. Weed Management and Tall Fescue Quality as Influenced by Mowing, Nitrogen, and Herbicides, *Crop Science* 33:1055–1061. Crop Science Society of America, Madison, WI.

Dernoeden, P.H. 2000. *Creeping Bentgrass Management: Summer Stresses, Weeds and Selected Maladies,* Ann Arbor Press, Chelsea, MI.

Derridj, S., V. Gregoire, J.P. Boutin, and V. Fiala. 1989. Plant Growth Stages in the Interspecific Oviposition Preference of the European Corn Borer and Relations with Chemicals Present on the Leaf Surfaces, *Entomol. Exp. Appl.* 53:267–276,

Dest, W.M. 1995. Turfgrass Clipping Management, *Turf Notes*, Volume 5, No. 2, UMass Extension, Worcester, MA.

Dest, W.M., S.C. Albin, and K. Guillard. 1992. Turfgrass Clipping Management, *Rutgers Turfgrass Proceedings 1992*. Rutgers University, New Brunswick, NJ.

DiMascio, J.A., P.M. Sweeney, T.K. Dannegerger, and J.C. Kamalay. 1994. Analysis of Heat Shock Response in Perennial Ryegrass Using Maize Heat Shock Protein Clones, *Crop Science* 34:798–804. Crop Science Society of America, Madison, WI.

Dinelli, D.F. 1997. IPM on Golf Courses, *IPM Practitioner*, Volume 19, No. 4, Bio-Integral Resource Center, Berkeley, CA.

Dinelli, D.F. 1999. Using Composts to Improve Turf Ecology, *USGA Green Section Record*, July/August 1999, USGA, Far Hills, NJ.

Dinelli, D.F. 2000. Composts to Improve Turf Ecology, *IPM Practitioner,* Volume 22, No. 10, Bio-Integral Resource Center, Berkeley, CA.

Dodson, R. 1996. Integrated Pest Management for Land Managers, *USGA Green Section Record*, Volume 34, No. 5, USGA, Far Hills, NJ.

DTN Weather Center, 11400 Rupp Drive, Burnsville, MN 55337, phone: 952-882-4337, fax: 952-882-4500, web site: www.dtnonline.com.

Edwards, C.A. and P.J. Bohlen. 1996. *Biology and Ecology of Earthworms:* 3rd ed. Chapman & Hall, London.

Elam, P. 1994. Earthworms: We Need Attitude Adjustment, *Landscape Management,* July, 1994. Advanstar Communications, Cleveland, OH.

Elliott, M.L. 2000. Black Box Research: Seeking Answers to the Effectiveness of Bacterial Inoculants, , *USGA Green Section Record*, Volume 38, No. 6, USGA, Far Hills, NJ.

Engage Agro Corporation, 315 Woodlawn Road West, Quelph, Ontario, Canada, N1H 7K8, phone: 800-900-5487, www.engageagro.com.

EPA 2000. http://www.epa.gov/pesticides/biopesticides/.

Ernst, D. 1995. *The Farmer's Earthworm Handbook*, Lessiter Publications, Brookfield, WI.

Esnard, J. 2000. Why Turf Needs Good Nematodes, *Cornell University Turfgrass Times*, Volume 11, No. 2, Cornell University, Ithaca, NY.

European Turfgrass Laboratories Limited, Unit 58, Stirling Enterprise Park, Stirling FK7 7RP Scotland, phone: 44-1786-449195, fax: 44-1786-449688.

Field Turf, Inc., 5050 Pare Street, Suite 280, Montreal, Quebec, Canada, H4P 1P3, phone: 800-724-2969, 514-340-9311, fax: 514-340-9374, www.fieldturf.com.

Foth, H.D. and B.G. Ellis. 1988. *Soil Fertility*, John Wiley and Sons, New York, NY.

Frank, J. 2000. Quantum Physics and Horticulture. Keynote Address, 2000 Long Island Organic Horticulture (LIOHA) Conference, LIOHA, Massapequa, NY.

Franklin, S. 1988. *Building a Healthy Lawn: A Safe and Natural Approach,* Storey Communications, Inc., Pownal, VT.

Fream, R.W. 2001. Do You Have Green Creep?, *USGA Green Section Record*, Volume 39, No. 1, USGA, Far Hills, NJ.

Fresh Aire Implements (Duke Merrion), 2283 South Rd. 300, East Danville, IN 46122, phone: 765-498-1070

Garlic Research Labs, 624 Ruberta Ave., Glendale, CA 91201, 800-424-7990, garlik@earthlink.net.

Gaugler, R. and H.K. Kaya. 1990. *Entomopathogenic Nematodes in Biological Control*, CRC Press, Boca Raton, FL.

Gaugler, R. 2000. Matching Nematodes to Target Pests, *IPM Practitioner*, Volume 22, No. 2, Bio-Integral Resource Center, Berkeley, CA.

Giesler, L.J., G.Y. Yuen, and G.l. Horst. 2000. Canopy Microenvironments and Applied Bacteria Population Dynamics in Shaded Tall Fescue, *Crop Science* 40:1325–1331, Crop Science Society of America, Madison, WI.

Gobran, G.R., W.W. Wenzel, and E. Lombi. 2001. *Trace Elements in the Rhizosphere*, CRC Press, Boca Raton, FL.

Grainge, M. and S. Ahmed. 1988. *Handbook of Plants with Pest Control Properties*, John Wiley & Sons. New York, NY.

Grant, J., M. Villani, and J. Nyrop. 1994. Predicting Grub Populations in Home Lawns, *Cornell University Turf Times,* Volume 5, No. 2. Cornell University, Ithaca, NY.

Grossman, J. 2000. Compost Cures: Microbial-Powered Composts and Compost Teas Are Becoming Respected Weapons in the War Against Pests, *Growing Edge*, Volume 11, No. 3, Corvallis, OR.

Growing Solutions, Inc., 160 Madison St., Eugene, OR 97402, phone: 541-343-8727 fax: 541-343-8374, e-mail: info@growingsolutions.com, web site: www.growingsolutions.com.

Hach Company, Products for Analysis, P.O. Box 389, Loveland, CO 80539, phone: 800-227-4224, fax: 970-669-2932, e-mail: orders@hach.com, web site: www.hach.com.

Hale, M.G. and D.M. Orcutt. 1987. *The Physiology of Plants Under Stress*, John Wiley and Sons, New York, NY.

Hall, R. 1994. Turf Pros Respond to Biostimulants, *Landscape Management*, Oct. 94, Advanstar Communications, Duluth, MN.

Hanna Instruments, Inc. 584 Park East Drive, Woonsocket, RI 02895-0849, phone: 800-426-6287 or 401-765-7500, fax: 401-765-7575, e-mail: sales@hannainst.com, web site: www.hannainst.com.

Harban, W.S. 1921. Winter Work on the Golf Course, *USGA Green Section Record*, Volume 1, No. 2, USGA, Far Hills, NJ.

Harper, J.C., II. 2001. Growing Turf Under Shaded Conditions, *Turf Magazine,* April 2001, Moose River Publishing, LLC, St. Johnsbury, VT.

Hartwiger, C. and P. O'Brien. 2001. Core Aeration by the Numbers, *USGA Green Section Record*, Volume 39, No. 4, USGA, Far Hills, NJ.

Hartwiger, C.E., C.H. Peacock, J.M. DiPaola, and D. K. Cassel. 2001. Impact of Light-Weight Rolling on Putting Green Performance, *Crop Science* 41:1179–1184, Crop Science Society of America, Madison, WI.

Heinrichs, E.A. Ed. 1988. *Plant Stress—Insect Interactions*, John Wiley & Sons, Inc., New York, NY.

Hendrix, P.F. 1995, *Earthworm Ecology and Biogeography*, CRC Press, Boca Raton, FL.

Herbert Ranch, Inc. (Pat Herbert), P.O. Box 65, Hollister, CA, 95024-0065, phone: 831-637-5515, fax: 831-637-0139, e-mail: patti@herbertranch.com, website: www.herbertranch.com

Hill, N.J., J.R. Heckman, B.B. Clarke, and J.A. Murphy. 1999. Take-all Patch Suppression in Creeping Bentgrass with Manganese and Copper, *Hort. Science* 34:891–892.

Hoitink, H.A.J. and M.E. Grebus. 1994. Nurseries Find New Value in Composted Products, *Biocycle*, The Journal of Composting and Recycling, 35:51–52, Emmaus, PA.

Hoitink, H.A.J. and A.G. Stone. 1995. Factors Affecting Suppressiveness of Composts to Plant Disease, *Hort. Science,* Volume 30, No. 4.

Hoitink, Harry A.J., A.G. Stone, and D.Y. Han. 1997. Suppression of Plant Diseases by Composts, *Hort. Science,* Volume 32, No. 2.

Huang, B. 2000. Summer Decline of Cool Season Turfgrasses: Heat Stress and Cultural Management, *Turfgrass Trends*, Volume 8, No. 6, Washington, DC.

Huang, B. and H. Gao. 2000. Growth and Carbohydrate Metabolism of Creeping Bentgrass Cultivars in Response to Increasing Temperatures, *American Society of Agronomy, Inc.*, 40, #4:1115–1120, Crop Science Society of America, Inc., Madison, WI.

Huang, P.M. and M. Schnitzer. 1986. *Interactions of Soil Minerals with Natural Organics and Microbes*, Soil Science Society of America, Inc., Madison, WI.

Huffaker, C.B. and R.L. Rabb. 1984. *Ecological Entomology*, John Wiley & Sons, Inc., New York, NY.

Hull, R.J. 1996. Nitrogen Usage by Turfgrass, *Turfgrass Trends*, Volume 5, No. 11, Washington, DC.

Hull, R.J. and J.T. Bushoven. 2001. A Metabolic Approach: Improving Nitrogen Use Efficiency in Turfgrasses, *USGA Green Section Record*, Volume 39, No. 4, USGA, Far Hills, NJ.

Hummel, N.W. Jr. 1993. Annual Bluegrass Biology and Control, *Cornell University Turfgrass Times* (CUTT), Volume 4, No. 1, Cornell University, Ithaca, NY.

Ingham, E.R. and M. Alms. 1999. *Compost Tea Manual 1.1*, Soil Food Web, Inc./Growing Solutions, Inc./ Ardeo, Inc., Eugene, OR.

ISTRC New Mix Lab LLC, 1530 Kansas City Road, Suite 110, Olathe, KS 66061, phone: 800-362-8873, fax: 913-829-8873.

Jackson, N. 1994. Winter Turf Problems, *Turf Notes*, November/December 1994. (Reprinted from *The Yankee Nursery Quarterly*, Volume 4, No. 4, 1994) New England Cooperative Extension Systems, University of Massachusetts, Worcester, MA.

Jenny, H. 1941. *Factors of Soil Formation*, McGraw–Hill Book Co., New York, NY.

Jiang, Y. and B. Huang. 2000. Effects of Drought of Heat Stress Alone and in Combination on Kentucky Bluegrass, *Crop Science* 40:1358–1362, Crop Science Society of America, Madison, WI.

Jiang, Y. and B. Huang. 2001. Osmotic Adjustment and Root Growth Associated with Drought Preconditioning-Enhanced Heat Tolerance in Kentucky Bluegrass, *Crop Science* 41:1168–1173, Crop Science Society of America, Madison, WI.

Jones, J. B. Jr. 1998. *Plant Nutrition Manual*, CRC Press, Boca Raton, FL.

Josephine Porter Institute of Applied Biodynamics, P.O. Box 133, Woolwine, VA 24185, phone: 540-930-2463, http://igg.com/bdnow/jpi/.

Kauffman, S. 2000. Pesticide Use Rises Slightly, *Golf Week's Superintendent News*, Volume 2, No. 9, Orlando, FL.

Keefer, R. F. 2000. *Handbook of Soils for Landscape Architects*, Oxford University Press, New York, NY.

Kenna, M. 2001. Nature Will Find a Way: Common Myths about Soil Microbiology, *USGA Green Section Record,* Volume 39, No. 3, USGA, Far Hills, NJ.

Kim, T.J., F. Rossi, and J. Neal. 1997. Ecological Aspects of Crabgrass Infestation in Cool-Season Turf, *Cornell University Turfgrass Times* (CUTT), Volume 8, No. 2, Cornell University, Ithaca, NY.

Koch, G. W. and H. A. Mooney, Eds. 1996. *Carbon Dioxide and Terrestrial Ecosystems*, Academic Press, Inc., San Diego, CA.

Kuo, Y., T.W. Fermanian, and D. J. Wehner. 2000. Nitrogen Utilization in Creeping Bentgrass, *Journal of Turfgrass Management*, Volume 3, No. 2, Haworth Press, Inc., Binghamton, NY.

Lamboy, J. 2000. Humus and Turf: Its Mysterious Role and Composition, *Turf Magazine*, February 2000, Moose River Publishing, LLC, St. Johnsbury, VT.

Lee, K.E. 1985. *Earthworms, Their Ecology and Relationships with Soils and Land Use,* Academic Press, Inc., Orlando, FL.

Leius, K. 1967. Influence of Wildflowers on Parasitism of Tent Caterpillar and Codling Moth, *Can. Entomol.* 99, pp. 444–446.

Lennert, L. 1990. The Role of Iron in Turfgrass Management in Wisconsin, USGA TGIF #:17611.

Leslie, A.R. Ed. 1994. *Handbook of Integrated Pest Management for Turf and Ornamentals*, Lewis Publishers, Boca Raton, FL.

Li, D., Y.K. Joo, N. E. Christians, and D.D. Minner. 2000. Inorganic Soil Amendment Effects on Sand-Based Sports Turf Media, *American Society of Agronomy, Inc.*, Volume 40, No. 4, Crop Science Society of America, Inc., Madison, WI.

Lilly, S. 1999. *Golf Course Tree Management,* Ann Arbor Press, Chelsea, MI.

Links Analytical, 22170 S. Saling Road, Estacada, OR 97023, phone: 503-630-7769, fax: 503-630-7764, e-mail: links@europa.com.

Lizzi, Y., C. Coulomb, C. Polian, P.J. Coulomb, and P.O. Coulomb. 1998. Seaweed and Mildew: What Does the Future Hold?, *Phytoma, The Defense of Plant,* No. 508:29–30, September, 1998.

Lloyd, J., Ed. 1997. *Plant Health Care: for Woody Ornamentals*, International Society of Arboriculture, Savoy, IL.

Lucas, R.E. and M.L. Vitosh. 1978. Soil Organic Matter Dynamics, Michigan State Univ. Research Report 32.91, Nov. 1978, East Lansing, MI.

Lyceum, P.O. Box 254, 18 Fairview St. (Westhampton Beach), Westhampton, NY 11977, phone: 631-288-2834, e-mail: jtf4647@aol.com, website: www.greenguerrilla.com.

Magdoff, F. and H. van Es. 2000. *Building Soils for Better Crops,* 2nd ed, Sustainable Agriculture Network, Beltsville, MD.

Makarov, I.B. 1986. Seasonal Dynamics of Soil Humus Content, *Moscow University Soil Science Bulletin,* Volume 41, No. 3: 19–26.

Martens, M-H.R. 2000. The Soil Food Web: Tuning in to the World Beneath Our Feet, *Acres USA,*Volume 30, No. 4, Austin, TX.

Maske, C. 2000. Personal communication. Maske's Organic Gardening, Decatur, IL.

Mason, J.R., L. Clark, and T.P. Miller. Evaluation of a Pelleted Bait Containing Methyl Anthranilate as a Bird Repellent, U.S. Department of Agriculture, Animal and Plant Health Inspection Service, Denver Wildlife Research Center, c/o Monell Chemical Senses Center, Philadelphia, PA.

McCarty, L.B., J.W. Everest, D.W. Hall, T.R. Murphy, and F. Yelverton. 2000. *Color Atlas of Turfgrass Weeds,* Ann Arbor Press, Chelsea, MI.

McKewon, M. 2001. Why Are Forecasts Always Wrong?, *Turf Magazine*, May 2001, Moose River Publishing, LLC, St. Johnsbury, VT.

Midwest Bio-Systems, 28933-35 E St., Tampico, IL 61283, phone: 800-335-8501, fax: 815-438-7028

Musser, H. B. 1962. *Turf Management*, McGraw-Hill, New York, NY.

Mycogen Corporation, 5501 Oberlin Drive, San Diego, CA 92121, phone: 800-745-7476, fax: 619-552-0459, www.mycogen.com.

Nakasaki, K., S. Hiraoka, and H. Nagata, 1998. A New Operation for Producing Disease-Suppressive Compost from Grass Clippings, *Applied and Environmental Microbiology*, Volume 64, No. 10: 4015–4020, American Society for Microbiology, Washington, D.C.

Neal, J. 1990. Waging War on Crabgrass, *Cornell University Turfgrass Times* (CUTT),Volume 1, No.1, Cornell University, Ithaca, NY.

Neal, J.C. 1992. Plan Before You Plant, *WeedFacts*, July 1992, Cornell University, Ithaca, NY.

Neal, J.C. 1993. Turfgrass Weed Management—An IPM Approach, *WeedFacts,* August 1993, Cornell University, Ithaca, NY.

Nektarios, P.A., T.S. Steenhuis, A.M. Petrovic, and J.Y. Parlange. 1999. Fingered Flow in Laboratory Golf Putting Greens, *Journal of Turfgrass Management*, Volume 3, No. 1, p. 53, Haworth Press, Binghamton, NY.

Nelson, E.B. Impacts of Conventional Turfgrass Pesticides on the Efficacy of Composted Amendments Used for the Biological Control of Turfgrass Diseases, unpublished, available through Cornell University, Ithaca, NY.

Nelson, E.B. Biological Control of Turfgrass Diseases, unpublished, available through Cornell University, Ithaca, NY.

Nelson, E.B. 1994. More Than Meets the Eye: The Microbiology of Turfgrass Soils, *Turf Grass Trends,* Volume 3, No. 2, Washington, DC.

Nelson, E.B. 1995. The Microbiology of Turfgrass Soils, *Cornell University Turfgrass Times* (CUTT), Volume 5, No. 4, Cornell University, Ithaca, NY.

Nelson, E.B. 1997. Biological Control of Turfgrass Diseases, *Golf Course Management*, July 1997, Lawrence, KS.

Nelson, E.B. 2001. What Are Those Microbes?, *Cornell University Turfgrass Times*, Volume 12, No. 1, Cornell University, Ithaca, NY.

Nelson, E.B. and F.S. Rossi. 2001. Creeping Bentgrass Cultivar Influences Biocontrol, *Cornell University Turfgrass Times*, Volume 12, No. 2, Cornell University, Ithaca, NY.

Nelson, M. 1997. Natural Areas: Establishing Natural Areas on the Golf Course, *USGA Green Section Record,* Volume 35, No. 6, USGA, Far Hills, NJ.

Neumann, S. and G.J. Boland. 1999. Influence of Selected Adjuvants on Disease Severity by Phoma Herbarium on Dandelion, *Weed Technology*, 13:675–679.

Norrie, J. and D.A. Hiltz. 1999. Seaweed Extract Research and Applications in Agriculture, March/April 1999, Agro-Food Industry Hi-Tech.

North American Kelp, 41 Cross St., Waldoboro, ME 04572, toll free: 888-662-5357, phone: 207-832-7506, fax: 207-832-6905, e-mail: nak@noamkelp.com, web site: www.noamkelp.com.

Northwest Irrigation, 30373 Hwy. 34, Albany, OR 97321, phone: 866-4-BREWER, or 541-928-0114, fax: 541-928-0307, e-mail: info@microbbrewer.com, web site: www.microbbrewer.com.

NRAES. 1992. *On Farm Composting Handbook,* Northeast Regional Engineering Service #54, Cornell Cooperative Extension, Ithaca, NY.

N.W. Hummel & Co., 35 King St., P.O. Box 606, Trumansburg, NY 14886, phone: 607-387-5694, fax: 607-387-9499, e-mail: SOILDR1@Epix.net, web site: www.turfdoctor.com.

O'Brien, P.M. 1996. The Magic of Sulfur, Volume 34 No. 3, *USGA Green Section Record*, USGA, Far Hills, NJ.

O'Brien, P.M. 1996. Optimizing the Turfgrass Canopy Environment with Fans, *USGA Green Section Record*, Volume 34, No. 4, USGA, Far Hills, NJ.

Olkowski, W., S. Daar, and H. Olkowski. 1991. *Common-Sense Pest Control*, The Taunton Press, Newtown, CT.

Peak Minerals – Azomite, Inc., P.O. Box 6588, Branson, MO 65615, toll free: 877-296-6483 417-334-8500, fax: 417-334-8825, e-mail: azomite@aol.com, web site: www.azomite.com.

Petrik Laboratories, 109 Harter Avenue, Woodland, CA 95776, phone: 530-666-1157, fax: 530-661-0489, website: www.petrik.com, e-mail: mike@petriklabs.com.

Petrovic, M.A., C.A. Sanchiroco, D.J. Lisk, R.G. Young, and P. Larrson-Kovach. 1993. Pesticide Leaching from Simulated Golf Course Fairways, *Cornell University Turfgrass Times* (CUTT), Volume 4, No. 2, Cornell University, Ithaca, NY.

Petrovic, M.A. 1997. The Art and Science of Turfgrass Soil Management, *Cornell University Turfgrass Times* (CUTT), Volume 11, No. 3, Cornell University, Ithaca, NY.

Petrovic, M.A. 2000. Oxygen Injection and Bentgrass Rooting, *Cornell University Turfgrass Times* (CUTT), Volume 11, No. 3, Cornell University, Ithaca, NY.

Petrovic, M.A. 2001. Understanding the Exchange: Turfgrass Nutrient Management and Cation Exchange Capacity, *Cornell University Turfgrass Times*, Volume 12, No. 1, Cornell University, Ithaca, NY.

Pfeiffer, E.E. *Weeds and What They Tell,* (reprinted by) Bio-Dynamic Literature, Wyoming, RI.

Phelan, P.L., J.F. Mason, and B.R. Stinner. 1995. Soil-Fertility Management and Host Preference by European Corn Borer, *Ostrinia nubilalis* (Hübner), on *Zeo mays* L.: A Comparison of Organic and Conventional Chemical Farming, *Agriculture, Ecosystems and Environment,* 56 (1995) 1–8 Elsevier Science, Inc., NY.

Phelan, P.L., K.H. Norris, and J.F. Mason. 1996. Soil-Management History and Host Preference by *Ostrinia nubilalis*: Evidence for Plant Mineral Balance Mediating Insect-Plant Interactions, *Environmental Entomology,* 25(6):1329–1336, Entomological Society of America, Lanham, MD.

Phelan, P.L. 1997. Soil-Management History and the Role of Plant Mineral Balance as a Determinant of Maize Susceptibility to the European Corn Borer, *Entomological Research in Organic Agriculture,* 1997:25–34.

Pierzynski, G. M., J. T. Sims, and G. F. Vance. 1994. *Soils and Environmental Quality,* Lewis Publishers, Boca Raton, FL.

Piper, C.V. and R.A. Oakley. 1921. Humus Producing Materials and the Making and Use of Compost, *USGA Green Section Record,* Volume 1, No. 4, USGA, Far Hills, NJ.

Piper, C.V. and R.A. Oakley. 1921. Some Suggestions for Fall Treatment of Putting Greens, *USGA Green Section Record,* Volume 1, No. 8, USGA, Far Hills, NJ.

Pike Lab Supplies (Bob Pike), RFD#2, Box 92, Strong, ME 04983, phone: 207-684-5131, e-mail: pike@inetme.com, website: www.maine.com/tse/pals/welcome.html

Players Turf International, L.L.C., 809 W. Detweiller Drive, Peoria, IL 61615, phone: 877-307-5530, fax: 309-693-0354, web site: www.playersturf.com, contact: Ray Sever.

Potter, D.A. 1998. *Destructive Turfgrass Insects: Biology, Diagnosis, and Control,* Ann Arbor Press, Chelsea, MI.

Prakash, A. and J. Rao. 1997. *Botanical Pesticides in Agriculture,* CRC Press, Boca Raton, FL.

Price, P.W., T.M. Lewinsohn, G.W. Fernandes, and W.W. Benson, Eds. 1991. *Plant–Animal Interactions,* John Wiley & Sons, Inc., New York, NY.

Quarles, W. 1996. New Microbial Pesticides for IPM, *IPM Practitioner,* Volume 18, No. 8, Berkeley, CA.

Quarles, W., Ed. 2001. Golf Course Pests, *IPM Practitioner* (Conference Notes), Volume 23, Nos. 5/6, Berkeley, CA.

Quarles, W., Ed. 2001. Black Cutworm IPM, *IPM Practitioner* (Conference Notes), Volume 23, Nos. 5/6, Berkeley, CA.

Rechcigl, J.E., Ed. 1995. *Soil Amendments: Impacts on Biotic Systems,* Lewis Publishers, Boca Raton, FL.

Ridzon, L. and C. Walters Jr. 1990. The Carbon Connection, *Acres USA,* Austin, TX.

Roberts, E. 1992. Private communication, The Lawn Institute, Pleasant Hill, TN.

Roberts, J. 1995. Spring Fertilization Jump Starts Turf, *Landscape Management,* February 1995, Advanstar Communications, Cleveland, OH.

Rossi, F.S. 1999. Everything You Ever Wanted to Know about Crabgrass . . . But Didn't Know Who to Ask, *Cornell University Turfgrass Times,* Volume 10, No. 4, Cornell University, Ithaca, NY.

Rossi, F.S. 2000a. Wondering about Golf Course Ecology, *Cornell University Turfgrass Times,* Volume 10, No. 3, Cornell University, Ithaca, NY.

Rossi, F.S. 2000b. Does Bentgrass Overseeding Work?, *Cornell University Turfgrass Times,* Volume 10, No. 4, Cornell University, Ithaca, NY.

Rossi, F.S. 2000c. Core Cultivation: A Necessary Evil?, *Cornell University Turfgrass Times,* Volume 11, No. 3, Cornell University, Ithaca, NY.

Rossi, F. S. 2001. Lessons from the Lorax: The Golf Industry Would Do Well to Pay Attention to Sage Advice from Dr. Seuss, *USGA Green Section Record*, Volume 39, No. 3, USGA, Far Hills, NJ.

Roth, S. 2001. A New Nematode Joins the Fight: Biological Control for Mole Crickets, *Turf Magazine,* March 2001, Moose River Publishing, LLC, St. Johnsbury, VT.

Sachs, P.D. 1999. *Edaphos: Dynamics of a Natural Soil System,* 2nd ed., Edaphic Press, Newbury, VT.

Sachs, P.D. 1996. *Handbook of Successful Ecological Lawn Care*, Edaphic Press, Newbury, VT.

Schmaderer, J. 2000. Rethinking the Rootzone: The Agronomic Benefits of Porous Ceramic Amendments, *Turf Magazine*, June 2000, Moose River Publishing, LLC, St. Johnsbury, VT.

Schmidt, R.E. and X. Zhang. Influence of Seaweed on Growth and Stress Tolerance of Grasses, T4.6.1, Dept. of Crop and Soil Environmental Sciences, Virginia Polytechnic Institute and State University, Blacksburg, VA.

Schmidt, R.E. and X. Zhang. 1998. Manipulation of Mineral Nutrition with Organic Growth Regulators on Turfgrass Antioxidant Activity and Tolerance to Environmental Stress, Dept. of Crop and Soil Environmental Sciences, Virginia Polytechnic Institute and State University, Blacksburg, VA.

Schmidt, R.E. and X. Zhang. 1998. Manipulation of Mineral Nutrition with Organic Growth Regulators on Creeping Bentgrass Turf and Tolerance to Environmental Stress, Dept. of Crop and Soil Environmental Sciences, Virginia Polytechnic Institute and State University, Blacksburg, VA.

Schriefer, D.L. 2000. Agriculture in Transition, *Acres USA,* Austin, TX.

Schriefer, D.L. 2000. From the Soil Up, *Acres USA*, Austin, TX.

Schultz, W. 1989. *The Chemical Free Lawn*, Rodale Press, Emmaus, PA.

Schumann, G.L. 1994. Predicting Rhizoctonia Blight (Hot Weather Brown Patch), *Turf Notes* Volume 4, No. 5, UMass Extension, Worcester, MA.

Schumann, G.L. 1994. Disease Control in Cool Season Grasses, *Landscape Management*, May 1994, Advanstar Communications, Cleveland, OH.

Schumann, G.L. 1995. Remember Last Summer, *Turf Notes*, Volume 5, No. 3, UMass Extension, Worcester, MA.

Schumann, G.L., P.J. Vittum, M.L. Elliott, and P.P. Cobb. 1998, *IPM Handbook for Golf Courses*, Ann Arbor Press, Chelsea, MI.

Senesac, A. 1992. Fall Weed Control, *Cornell University Turfgrass Times* (CUTT), Volume 3, No. 3, Cornell University, Ithaca, NY.

Senn, T.L. and A.R. Kingman. 1973. A Review of Humus and Humic Acids, Clemson Univ. Research Series #145, March 1, 1973, Clemson, SC.

Senn, T.L. 1987. *Seaweed and Plant Growth*, Department of Horticulture, Clemson University, Clemson, SC.

SFI 2000. Soil Food Web, Inc., Corvallis, OR 97330, www.soilfoodweb.com.

SFI 2000a, http://www.soilfoodweb.com/products/bacterialresources.html.

Simpson, T.C. 1931. *The Game of Golf: The Upkeep of a Golf Course,* pp. 197–218, J. B. Lippincott Co., Philadelphia, PA.

Smigocki, A., S. Heu, I. McCanna, C. Wozniak, and G. Buta. Insecticidal Compounds Induced by Regulated Overproduction of Cytokinins in Transgenic Plants, in: *Advances in Insect Control*, pp. 225–236, N. Carozzi and M. Koziel, Eds. 1997, Taylor & Francis Ltd., Washington, DC.

Smith, C.M. 1989. *Plant Resistance to Insects: A Fundamental Approach*, John Wiley & Son, Inc., New York, NY.

Spectrum Technologies, Inc., 23839 West Andrew Road, Plainfield, IL 60544, phone: 815-436-4440, fax: 815-436-4460, e-mail: specmeters@aol.com, web site: www.specmeters.com.

SSSA# 19. 1987. *Soil Fertility and Organic Matter as Critical Components of Production Systems*. Soil Science Society of America, Inc., Madison, WI.

Stevenson, F. J. 1986. *Cycles of Soil: C, N, P, S, Micronutrients*, John Wiley and Sons, New York, NY.

St. Gabriel Labs, 14540 John Marshall Highway, Gainsville, VA 20155, phone: 703-754-3823.

Stuart, K. 1992. A Life with the Soil, *Orion*, Volume 11, No. 2, Spring 1992, pp. 17–29, Myrin Institute, New York, NY.

Sullivan, W. M., Z. Jiang, and R. J. Hull. 2000. Root Morphology and Its Relationship with Nitrate Uptake in Kentucky Bluegrass, *Crop Science Journal*, Volume 40, No. 3, CSSA, Madison, WI.

Summa Minerals, 2310 Sherman Place, Las Vegas, NV 89102, phone: 702-256-2990, e-mail: summaminerals@mail.com.

Sumner, M.E., Ed. 2000. *Handbook of Soil Science*, CRC Press, Boca Raton, FL.

Tani, T. and J.B. Beard. 1997. *Color Atlas of Turfgrass Diseases: Disease Characteristics and Control,* Ann Arbor Press, Chelsea, MI.

Terry, L.A., D.A. Potter, and P.G. Spicer. 1993. Insecticides Affect Predatory Arthropods and Predation of Japanese Beetle (Coleoptera: Scrabaeidae) Eggs and Fall Armyworm (Lepidoptera: Noctuidae) Pupae in Turfgrass, *Journal of Economic Entomology*, Volume 86, No. 3.

Thomas Turf Services, Inc., 2151 Harvey Mitchell Parkway South, Suite 302, College Station, TX 77840-5247, phone: 979-764-2050, fax: 979-764-2151, e-mail: soiltest@thomasturf.com.

Thurn, M. 1995. Composts as Soil Amendments, *Cornell University Turfgrass Times* (CUTT), Volume 5, No. 4, Cornell University, Ithaca, NY.

Tifton Physical Soil Testing Laboratory, Inc., 1412 Murray Avenue, Tifton, GA 31794, phone: 912-382-7292, fax: 912-382-7992, e-mail: pgaines@surfsouth.com.

Torello, W.A., H. Gunner, and M. Coler. 1998. Microbial Population Dynamics in Sand Green Profiles: The Effects of Vesicular-Arbuscular Mycorrhiza Fungi on Turfgrass Vigor, Disease/Insect Activity, Environmental Stress Resistance and Fertilizer Efficiency, *1998 Turfgrass Field Day*, pp. 47–48, University of Massachusetts, Amherst, MA.

Torello, W.A., H. Gunner, and M. Coler. 1999. Biological Disease Control in Golf Turf: A Unique Approach Utilizing Newly Developed Carrier Technology for a New Anti-Pathogenic Activity Bacterium, *1999 Turfgrass Field Day*, p. 25, University of Massachusetts, Amherst, MA.

Toro Company, Commercial Division, 8111 Lyndale Avenue South, Bloomington, MN 55420, phone: 952-888-8801, www.toro.com.

Trenholm, L.E., R.N. Carrow, and R.R. Duncan. 2000. Mechanisms of Wear Tolerance in Seashore Paspalum and Bermudagrass, *Crop Science* 40:1350–1357, Crop Science Society of America, Madison, WI.

Turf Diagnostics & Design, Inc., 310A North Winchester St., Olathe, KS 66062, phone: 913-780-6725, fax: 913-789-6759, e-mail: sferro@turfdiag.com.

Turfgrass Information Center, Michigan State University Libraries, 100 Library, East Lansing, MI 48828-1048, phone: (517) 353-7209, fax: 517-353-1975, e-mail: tgif@msu.edu.

Turgeon, A.J., Ed. 1994. *Turf Weeds and Their Control*, American Society of Agronomy, Inc., Crop Science Society of America, Inc., Madison, WI.

Unruh, J.B., N.E. Christians, and H.T. Horner. 1997. Herbicidal Effects of the Dipeptide Alaninyl-Slanine on Perennial Rygrass (*Lolium perenne L.*) Seedlings, *Crop Science Jour-*

nal, Volume 37, No. 1, pp. 208–212, American Society of Agronomy, Inc., Crop Science Society of America, Inc., Madison, WI.
U.S. Composting Council, P.O. Box 407, Amherst, OH 44001-0407, phone: 440-989-2748, website: www.compostingcouncil.org.
Uva, R.H., J.C. Neal, and J.M. DiTomaso. 1997. *Weeds of the Northeast*, Cornell University Press, Ithaca, NY.
van Veen, A. and P.J. Kuikman. 1990. Soil Structural Aspects of Decomposition of Organic Matter by Micro-organisms, *Biogeochemistry*, Dec. 1990, Volume 11, No. 3, pp. 213–233.
Vargas, J.M. Jr. 1994. *Management of Turfgrass Diseases*, 2nd ed., Lewis Publishers, Boca Raton, FL.
Vavrek, R.C. 1990. Beneficial Turfgrass Invertebrates, *USGA Green Section Record*, Nov./Dec. 1990, Volume 28, No. 6.
Vermeulen, P.H. 1997. Know When to Over-Irrigate: An Easy Way to Monitor Soil Salinity, *USGA Green Section Record*, Volume 35, No. 5, USGA, Far Hills, NJ.
Villani, M. 1998. Moisture Effects on Entomopathogenic Nematodes, *Cornell University Turfgrass Times* (CUTT), Volume 9, No. 1, Cornell University, Ithaca, NY.
Vilter, H., K.W. Glombitza, and A. Grawe. 1983. Peroxidases from Phaeophyceae: Extraction and Detection of the Peroxidases, *Botanica Marina,* Volume 26, pp. 331–340.
Vilter, H. 1983. Peroxidases from Phaeophyceae: Catalysis of Halogenation by Peroxidases from *Ascophyllum nodosum*, *Botanica Marina*. Volume 26, pp. 429–435.
Vittum, P.J. 1984. Effect of Lime Applications on Japanese Beetle (Coleoptera:Scarabaeidae) Grub Populations in Massachusetts Soils, *Journal of Economic Entomology,* Volume 77 pp. 687–690.
Vittum, P.J. 1994. Hyperodes Update, *Turf Notes*, Volume 4, No. 5, UMass Extension, Worcester, MA.
Vittum, P.J. 1995. Black Turfgrass Ataenius Update, *Turf Notes*, Volume 5, No. 2, UMass Extension, Worcester, MA.
Vittum, P.J. 1999. Identifying and Managing White Grubs in New England, *Turf Notes*, Volume 8, No. 2, UMass Extension, Worcester, MA.
Waisel, Y., A. Eshel, and U. Kafkafi, Eds. 1996. *Plant Roots: The Hidden Half*, 2nd ed., Marcel Dekker, Inc., New York, NY.
Waksman, S.A. 1936. *Humus*, Williams and Wilkins, Inc., Baltimore, MD.
Wallace, A. and R.E. Terry Eds. 1998. *Handbook of Soil Conditioners: Substances That Enhance the Physical Properties of Soil*, Marcel Dekker, Inc., New York, NY.
Walters, C. Jr. 1991. Weeds: Control Without Poisons, *Acres USA*, Kansas City, MO.
Watkins, J.E. 1996. Nitrogen Fertilization's Effect on Turfgrass Disease Injury, *Turfgrass Trends,* Volume 5, No. 11, Washington, DC.
Watschke, T.L., P.H. Dernoeden, and D.J. Shetlar. 1995. *Managing Turfgrass Pests*, Lewis Publishers, Boca Raton, FL.
White, P. 2000. Leave 'em Lie, *Turf Magazine*, September 2000, Moose River Publishing, LLC, St. Johnsbury, VT.
White, P. 2001a. Beam Me Up: The Future Is Now for One Company's New Mower, *Turf Magazine,* April 2001, Moose River Publishing, LLC, St. Johnsbury, VT.
White, P. 2001b. Artificial Flavors: A Number of New Products Have Put Artificial Turf Back in the Game, and Even on Some Lawns, *Turf Magazine*, August 2001, Moose River Publishing, LLC, St. Johnsbury, VT.
White, W.C. and Collins, D.N., Eds. 1982. *The Fertilizer Handbook,* The Fertilizer Institute, Washington, DC.
Wilkinson, R.E. Ed. 1994. *Plant-Environment Interactions*, Marcel Dekker, Inc., New York, NY.

Williams, N.D. and J.C. Neal. 1993. Annual Bluegrass Biology and Control, *Cornell University Turfgrass Times,* Volume 4, No. 1, Cornell University, Ithaca, NY.

Wiseman, B.R. and R.R. Duncan. 1996. Resistance of *Paspalum* spp. to *Spodoptera frugiperda* (J.E. Smith) (Lepidoptera: Noctuidae) Larvae, *Journal of Turfgrass Management,* Volume 1, No. 4, p. 23, Haworth Press, Binghamton, NY.

Witteveen, G. and M. Bavier. 1998. *Practical Golf Course Maintenance*, Ann Arbor Press, Chelsea, MI.

Woods End Research, POB 297, 1850 Old Rome Rd., Mt. Vernon, ME 04352, phone: 207-293-2457, fax: 207-293-2488, e-mail: solvita@woodsend.org, web: www.woodsend.org.

Xu, Q. and B. Huang. 2000. Growth and Physiological Responses of Creeping Bentgrass to Changes in Air and Soil Temperatures, *Crop Science,* Volume 40, pp. 1363–1367, Crop Science Society of America, Madison, WI.

Xu, Q and B. Huang. 2000. Effects of Differential Air and Soil Temperature on Carbohydrate Metabolism in Creeping Bentgrass, *Crop Science,* Volume 40, pp.1368–1374, Crop Science Society of America, Madison, WI.

Zhang, W., W. A. Dick, and H. A. J. Hoitink. Compost-Induced Systemic Acquired Resistance in Cucumber to Pythium Root Rot and Anthracnose, *Journal of the American Phytopathological Society*, Volume 86, pp. 1066–1070, St. Paul, MN.

Zhang, X. and R.E. Schmidt. 2000. Hormone-Containing Products' Impact on Antioxidant Status of Tall Fescue and Creeping Bentgrass Subjected to Drought, *Crop Science,* Volume 40, pp. 1344–1349, Crop Science Society of America, Madison, WI.

Zontek, S. 2000. Using Compost to Improve Poor Soils, *USGA Green Section Record,* May/June 2000, USGA, Far Hills, NJ.

Index

Page numbers in **bold** indicate illustrative material.